TRACE ELEMENTS IN AGRICULTURE

VINCENT SAUCHELLI
Consultant to the Fertilizer Industry

Formerly, Director of Agricultural Research, Davison Chemical Division, W. R. Grace & Co. and Chemical Technologist, National Plant Food Institute, Washington, D.C.

VAN NOSTRAND REINHOLD COMPANY
New York Cincinnati Toronto London Melbourne

Van Nostrand Reinhold Company Regional Offices:
New York, Cincinnati, Chicago, Millbrae, Dallas

Van Nostrand Reinhold Company Foreign Offices:
London, Toronto, Melbourne

Published by Van Nostrand Reinhold Company
450 West 33rd Street, New York, N.Y. 10001

Published simultaneously in Canada by
D. Van Nostrand Company (Canada), Ltd.

15 14 13 12 11 10 9 8 7 6 5 4 3 2 1

PREFACE

This book claims no originality. It is neither a critical nor an exhaustive survey of the literature. An author of this book type is necessarily a compiler, since he cannot hope to have made original contributions to the many areas included in a general text. I have used the published contributions of many specialists available in books, journal articles, and experiment station bulletins. It is not possible to give credit to all contributors, and I hope I have quoted accurately and without distortion of meaning. My sincere thanks are also given to those who kindly furnished me with illustrations of deficiency or toxic symptoms in plant and animals.

Trace elements, micronutrients, minor elements, oligoelements, or spürenelementes are synonymous terms and apply to a number of chemical elements that are required in extremely small amounts but that are essential for the normal life processes of living organisms. The literature of the trace elements has accumulated to a vast extent during the past 30 years. With the exception of 3 or 4 books on special aspects of the subject, the bulk of the literature is primarily in the form of bulletins, journal articles, and papers read at scientific society meetings. It seems timely to present a digest of this information in a convenient form and in as nontechnical jargon as possible. Many persons interested in agriculture and fertilizers do not have an easy access to the literature on this subject, nor would they have the time to look up and study the scattered information. Those persons in the fertilizer and animal feed industries particularly need practical, authoritative information on the mineral requirements of plants and animals, the toxic effects caused by an excess, or disease effects induced by an insufficiency of the various trace elements.

I have tried to explain a few aspects of this broad subject, and it will be obvious to the specialists that I have touched only very lightly on a small proportion of the topics that must occur to their minds.

My reading of the literature reveals that the great majority of persons in world agriculture use and understand the term "trace element." Originally this term was used by the chemical analyst to designate the presence of an element detectable qualitatively by his technique but indeterminable quantitatively because it was present in such insignificantly minute quantities. These were reported as being present in trace amounts, and hence the term "trace element." Some physiologists and agronomists are recommending that the term "micronutrient" be used. In Europe the terms oligoelement, mikro-element, spürenelemente and others are used to designate trace elements. I have preferred to use "trace element" in most cases, although I have also employed the term micronutrient synonymously.

This book is not intended to compete in any way with textbooks on animal and plant physiology and pathology.

Some overlapping and repetition occur between parts of the book. I have not tried to eliminate this situation. I have often wished to see, in my reading, one subject dealt with as a whole in one chapter. I dare say others have had the same desire, and this also applied to references placed as footnotes on the page of occurrence rather than grouped at the end of the chapter.

Plants constitute only one fraction of the total cycle of life, although a most important part. Animals also need trace elements for normal metabolic processes. If to the mineral nutritional requirements of higher plants are added those needed by animals and some species of "lower" plant life (e.g., fungi, bacteria, algae), the usual list of essential trace elements is increased by nine more, namely, cobalt, iodine, vanadium, selenium, sodium, fluorine, silicon, aluminum, chromium. The biologist faces the question whether to consider plants as primary foods, expected to secure the extra elements for the convenience of animals in addition to requiring them for their own life processes.

This book should serve the general reader seeking to keep abreast of the developments in this field, and should particularly interest and serve the fertilizer field representative who has to advise farmers and dealers, the vocational agricultural teacher, agronomy student, county agent, horticulturist, and those engaged in preparing and distributing animal feeds.

VINCENT SAUCHELLI

Baltimore, Maryland
August, 1969

CONTENTS

PREFACE / iii

GENERAL INTRODUCTION / 2

1 HISTORICAL BACKGROUND / 3

Water-culture Technique / 8

2 BIOPHISICOCHEMICAL RELATIONSHIPS / 10

Atoms; Molecules; Elements / 10   Essentiality of Defined Nutrient / 12
Solutions / 13   Ions; Emulsions; Colloidal Suspensions / 13

3 SOIL-PLANT RELATIONSHIPS / 15

Anions; Cations / 15   Green Leaves (Chlorophyll) / 16   Roots; How Plants
Feed / 16

4 TRACE ELEMENTS IN NUTRITION / 18

Historical Background / 18   The Trace Element Problem / 23   Trace Elements
and Soils / 24   Mineral Nutrition of Plants / 26   Genetics and Crop Yields / 29
Visual Diagnosis / 34

5 DEFICIENCY SYMPTOMS / 35

Modern Depletion Factors / 35

**PROPERTIES OF TRACE ELEMENTS; BIOCHEMICAL
RELATIONSHIPS IN PLANTS AND ANIMALS** / 38

6 IRON / 39

Fe in Soils / 40   Iron Deficiency / 45   Diagnostic Methods / 48   Corrective
Measures / 49   Chelating Agents / 52   How Chelating Agents Work / 53
How Chelates are Used / 53   Iron Salts Used / 54   Iron in Grassland Herbage /
54   Iron in Animal Diet / 55   Piglet Anemia / 55   Rickets / 56   For Further
Reading / 56

7 MANGANESE / 58

Chemical Relationships / 59 Determining the Amount of Mn / 60 Manganese in Soils / 61 Reactions in Soils / 63 Manganese: Plant Relationships / 64 Mn Deficiency in Soils / 65 Corrective Measures / 67 Mn Deficiency in Plants; Symptoms / 72 Toxicity Symptoms / 74 Manganese in Animal Metabolism / 74 Functions of Mn in Animal Body / 77 Toxic Dose of Mn / 77 Mn and Fertilizer / 79 General References / 80

8 BORON / 81

Boron: its Chemistry / 83 Deficiency in Soils and Crops / 86 Deficiency Symptoms / 87 Boron in Fertilizers / 88 Official Recommendations on Boron Use / 94 General Consideration / 98

9 ZINC / 107

Chemical and Physical Characteristics / 108 Solubility / 109 Sources for Fertilizer Uses / 109 Zinc in Plants / 110 Function of Zn in Plants / 111 Zinc in Soil / 114 Solubility Value / 116 Limiting Effects on Zinc Reactions in Soil / 118 Recent Reports from State Agriculture Experiment Stations / 119 Zinc in Grassland Herbage (Temperate Regions) / 123 Zinc Deficiency in Sugar Cane / 124 Zinc Deficiency in Tobacco / 125 Major Zn Defrciency Symptoms in Plants / 126 Zinc in Animals / 127 Deficiency Symptoms / 127 Conversion Chart / 132

10 MOLYBDENUM / 133

Chemical Relationships / 136 Mo in Soils / 136 Mo in Plants / 138 Function in Plants / 139 Mo Deficiency and Symptoms / 141 Toxicity to Animals / 143 Mo in Grassland Herbage / 144 Corrective Measures / 146 Some Recent Reports of Experiments Involving Mo / 147 General References / 149

11 COPPER / 151

Copper Chemistry / 152 Solubility of Copper Compounds in 100 Parts Water / 154 Copper in Soils / 154 Copper in Enzymes / 159 Copper in Green Plants / 160 Deficiency Symptoms / 161 Interrelationships / 165 Copper in Animals / 167 Functions of Cu / 167 "Teart" / 168 Other "Diseases" / 168 Copper as a Fungicide / 169 Comparative Toxicity / 169 Some General References / 170

12   CHLORINE / 172

Chemistry of Chlorine / 173   Sources of Chlorine / 174   Chlorine in Plants / 174   Deficiency Symptoms / 176   Chlorine in Animal Nutrition / 177

13   SODIUM / 180

List of Benefits / 181   Sodium and Soil Structure / 182   Sodium in Plants / 183 Sodium in Animals / 184   Sodium as Fertilizer / 184

14   SELENIUM / 186

Chemistry of Selenium / 188   Research and "Factor 3" / 188   Se Supplementation / 189   Treatment / 190   Selenium as Insecticide / 194   Health Hazards / 195   Analytical Methods / 195

15   COBALT / 197

Chemistry of Cobalt / 199   Cobalt Deficiency Disease / 199   Cobalt and Vitamin B / 200   Function of Cobalt / 201   Cobalt in Soils / 202   Cobalt in Plants / 202   Corrective Methods / 203   Fertilizing Pastures / 204   Drenching / 206   Salt Lick / 206

16   IODINE / 207

Iodine in Soils / 208   Iodine in Plants / 209   Iodine in Animal Nutrition / 210 Iodine in Animal Products / 211   Iodine and Goiter / 211

17   FLUORINE / 213

Fluorine Sources / 213   Fluorine in Plants / 214   Fluorine in Animal Nutrition / 215   Fluorosis / 215

18   NICKEL, LITHIUM, VANADIUM, SILICON, ALUMINUM / 217

Nickel / 217   Lithium / 218   Vanadium / 219   Silicon / 219   Silicon in Animals / 221   Aluminum / 221

FERTILIZER AND MARKETING PROBLEMS

19   TRACE ELEMENTS IN FERTILIZERS / 225

20   PRIME PRODUCERS AND MARKETING / 228

Trace Element Carriers / 228   Size of Market / 231

21 ANALYSIS BY ATOMIC ABSORPTION / 233

Colorado Soil Test for Zn and Fe / 234   Analysis of Fritted Nutrients / 234 References for Further Reading / 236

APPENDIX / 237

Micronutrient Deficiencies / 237   Boron / 238   Iron / 239   Manganese / 239 Zinc / 239   Molybdenum / 239   Copper / 239   Cobalt / 240   Selenium / 240 Aluminum / 240

INDEX / 242

TRACE ELEMENTS IN AGRICULTURE

GENERAL INTRODUCTION

1

HISTORICAL BACKGROUND

Despite mankind's centuries of acquaintance with crops and forests, the essential concepts of plant nutrition have evolved not more than a century ago. Early in the 19th century, evidence seemed to indicate that plants were composed of chemical elements derived from three sources, air, water, soil. The bulk of plant substance, usually about 95% of its dry matter, was made up of the three elements, oxygen, carbon, and water, derived primarily from air and water rather than from the soil. Central to this concept and quite revolutionary in thinking is the stellar role played by the element carbon in the metabolism of the plant. Science showed that synthesis was accomplished not by the roots, the intermediary organ of nutrition, but by the leaves and other green, aboveground parts of the plant, under the influence of light in the process now known as photosynthesis.

It is difficult to understand how some of the world's foremost thinkers, from Aristotle on down, could fail to comprehend the process of photosynthesis. This and related discoveries, however, could not have been made without the prior development of the experimental method and of the physicochemical sciences. Only by using the experimental approach, first outlined by Francis Bacon and strikingly employed by Galileo, was it possible to understand the laws of nature

3

and to apply them to interpret natural phenomena intelligently. Thus, the clue to the manner by which the green leaf's exposure to sunlight results in carbon accumulation was revealed. Knowledge of the photosynthetic process had to await Priestley's discovery in the late 18th century of oxygen and the remarkable demonstration by the ill-starred Lavoisier[1] of the composition of water and carbonic acid. Other valuable contributions to the elucidation of the complex photosynthetic process were subsequently made by such pioneers as Jan Ingenhousz[1], Jean Senebier[1], and Theodore de Saussure. By utilizing solar energy the green plant is able to synthesize a vast array of organic carbon compounds on which we depend for fossil fuels, proteins or their building blocks, the amino acids, sugars, and other carbohydrates, fats, vitamins, and many more products used for food, fiber, and shelter.

Photosynthesis is entitled to receive extended mention here because of its basic importance in our lives; ultimately life and civilization depend on it. Man and animals depend directly and indirectly on plants for food. Plants also supply the power employed both in the animal body and in chemical processing. The amount of plant products consumed, such as coal, petroleum, timber, is commonly considered an index of the degree of a country's industrial development. In addition to the food materials supplied by plants that are consumed by the animal body for its growth and maintenance, other materials, equally important, serve as natural preventive medicines. These are called vitamins, and if present in the diet, secure health and vigor, but, absent, result in malnutrition and disordered growth.

Plant nutrition, as presently understood, was for centuries enshrouded in obscurity and myths. A few Greek and Roman investigators recorded several empirical concepts about the art and practices of agriculture. Apparently they knew from experience and observation that somehow soils on which crops are continuously grown suffer a loss in fertility, and that applying lime, animal manures, and turning under green manures benefited the soil's productivity. Moreover, they believed plant food existed in the soil in a form directly absorbable by plants for their nourishment. Even as late as the middle of the 19th century, considerable confusion existed among eminent scientists of the period regarding the truth about plant nutrition and the role played in the process by soil and air constituents. Soil humus or some other substance present in it was considered the significant "principle." Lime and

[1] Joseph Priestley (1733–1804); Carl W. Scheele (1742–1786); Antoine L. Lavoisier (1743–1794); Jan Ingenhousz (1730–1799); Jean Senebier (1742–1809).

certain salts were recognized as playing a supplementary role to the major function of humus which was considered the source of the carbon taken up by plants. For example, A. D. Thaer[2], an influential leader of that period, asserted the following:

> "the fertility of the soil depends entirely on humus, since besides water, it alone furnishes in the soil, nutrients to plants. Without it no individual life is possible. Mineral fertilizers which do not contain any organic matter, act largely or entirely through the decomposition processes which they stimulate."

Other eminent contemporaries[3], Berzelius, Davy, Mulder, maintained that humus furnished an important supply of organic foods to crops. Such thinking was not too far from concepts held by Columella, the Roman, who in A.D. 60 published a handbook titled "Husbandry," which contained suggestions on farming that were followed for nearly 1,500 years. In the 16th century speculations on the subject flourished in many European countries, and some keen observations recorded in this period were confirmed years later by modern experimental methods. In France Palissy suggested[4], in 1563, that plowed under manures and plant residues returned to the soil many of those substances which had been removed in growth.

For centuries prior to the chemical age and the beginning of applied experimental methods, many completely mistaken or half-mystical ideas about plant nutrition were believed by natural philosophers of that period. One popular but fantastic concept was that plants absorbed their foods partly preformed "from the juices of the soil." Francis Bacon in the 17th century had suggested that the principal food of plants was water, that soil served mainly as anchorage for plants, and that each plant derived from the soil its own particular nourishment, and consequently when the same kind of plant was grown continuously on the same soil, it would deplete the reserves of that particular substance. Bacon's idea about water as a prime nutrient was dramatically upheld in a clever experiment conceived and conducted by Jan Baptiste von Helmont. He put 200 pounds of dry soil in a large container, watered the soil, and planted in it a willow cutting weighing 5 pounds. After a 5-year period during which only rain or distilled water was applied, the weight of the willow was determined as 163 lb 3 oz. He

[2] Thaer, A. D. 1809–1812. "Grundsätze der rationallen Landwirtschaft," Vol. 1–4, Berlin.
[3] J. J. Berzelius (1825–1831); H. Davy (1813); G. J. Mulder (1849).
[4] Palissy, B., "Recepte veritable," 1563.

could account for all but 2 ounces of the original 200 pounds of soil. Since nothing but water had been applied he concluded, falsely as we know now, that the principle of vegetation was water, that is to say, that plant substance was made from water alone. The loss of 2 ounces in the weight of the soil was explained as experimental error.

John Woodward[5], an English investigator of that period (1699), could not accept the validity of the conclusions drawn by von Helmont. He grew spearmint plants in water derived from different sources—rain, river, sewage, and other sewage to which garden mold had been added. From the results of his investigation he concluded that plants needed certain "terrestrial matter" for their growth, and water was not the exclusive "principle of vegetation."

Several similar experiments carried out during that period were forerunners, albeit in crude form, of the more precise, modern type of experiment for determining the inorganic requirements of plants indispensable to their growth.

The first recorded experiments in plant growth have previously been alluded to, namely, van Helmont's and Woodward's. The experiments were well conceived and excellently carried out. But van Helmont interpreted the result in an erroneous way; he is to be judged, however, by the facts and beliefs of his time. Using the experimental approach as a tool was frowned upon by the contemporary scholastic philosophers who believed that plants and animals were all derived from the four elements fire, earth, air, and water, and that the minerals in them occurred by a process of transmutation. What van Helmont proved, accordingly, was that one of the four elements, water, was the cause of plant growth. So strongly implanted was this opinion that it persisted for 200 years. The brilliant experiments of John Woodward, and of Joseph Priestley and Carl Scheele, who discovered oxygen in 1774, and found its relation to green plants, together with the contemporary findings of Ingenhousz and Senebier, were unable to alter this old point of view. These investigations had actually found the following essential facts about photosynthesis: that green plants absorb CO_2 from the air and liberate oxygen (O_2) in the light, that the reverse occurs in the dark, and that the amount of O_2 released is proportionate to the CO_2 used. But because of the dominant doctrine that all life must derive from one of the four basic elements, these otherwise brilliant discoveries missed what we now know are the central facts of plant physiology.

[5] Woodward, J., *Phil. Trans. Roy. Soc.*, London **21** (1699).

Not until the publication in 1804 of de Saussure's research[6] was the fallacy of transmutation routed from its dominance. Using the techniques and reasonings of the new chemistry established by Lavoisier, de Saussure was the first to offer incontrovertible facts and conclusions that much of the dry matter of plants was derived from water and from CO_2 of the air. He used considerable data from chemical analyses of the ash of different parts of plants to show that the mineral portion varied with the plant, the part analyzed, the season, and the soil on which it was grown. He deduced that the minerals in the ash had come from the soil and were not a product transmuted within the plant. Yet, the belief in the transmutation theory was not completely dispelled by these excellent efforts of de Saussure. The death blow was finally struck by an experiment by Wiegmann and Polstorff[7] in 1842 in which they grew plants from seed under strictly controlled conditions to prove that plants did not produce minerals within their tissues. Some details of this historical experiment are worth reporting. Seeds of the cress plant were grown in a platinum crucible, a porous mass of fine platinum wire being substituted for soil, and placed under a bell jar, through which they caused a gas mixture to circulate, consisting of 78 volumes of nitrogen, 21 volumes of oxygen, and 1 volume of carbon dioxide. Distilled water only was used to water the plants. Growth of the 28 germinated seeds under these controlled conditions was permitted to continue for 26 days, after which the plants were dried and ashed, yielding 0.0025 gm of ash. The ash from 28 seeds used as control also weighed 0.0025 gm. This equivalence in weight was a convincing demonstration that no new mineral matter was engendered within the plant. Thus the belief about transmutation was effectively exploded, and thereafter, the science of plant growth as we now know it had a chance to develop unhampered by traditional, repressive opinions. Our current belief is that the plant can generate all its complex chemical compounds and tissues from the chemical elements it absorbs from soil and air.

In the course of development certain lines of investigation were pursued to establish scientific data to buttress the new concept. First, it was necessary to discover which elements were essential for growth. Then, in turn, it was necessary to elaborate diagnostic techniques to do the following: (1) produce and describe deficiency symptoms; (2)

[6] Saussure, N. T. de, "Recherches chimiques sur la vegetation," Paris, 1804.

[7] Wiegmann, A. F. and Polstorff, L., "Über der anorganischen Bestandtheile der Pflanzen," Brunswick, 1842.

develop techniques for analyzing soils or plants for the purpose of determining the deficiency of an element or as a guide in assessing potential deficiencies; and (3) determine the part that essential elements play in the growth cycle of a plant.

Tools are necessary in any attempt to extract information or to build concepts and techniques for assessing potentialities. Such tools may be inadequate, as in van Helmont's investigation, or highly developed and adequate, as in modern scientific investigations. Today our scientists are equipped with concepts based on experimental and factual data, the physicochemical sciences, biochemical knowledge of the functions of enzymes, vitamins, and hormones, as well as skills in the use of radioisotopes as tracer elements, and exceptional instrumentation developed by an advanced technology.

Water-culture Technique

In 1860 the plant physiologists, J. Sachs[8] and W. Knop[9], improved the water-culture or soilless method, where plants are grown with roots immersed in a solution containing a specified number and amount of nutrient salts. They concluded that if the culture solution contained salts furnishing N, S, P, K, Ca, Mg, Fe, it was possible to grow a normal plant. Experience showed that this water-culture technique was very efficient for finding out the inorganic nutritional needs of plants. It has been used ever since as an important tool for the scientific elucidation of plant nutrition. Plants of many species have been successfully grown in water culture. D. R. Hoagland and D. I. Arnon in more recent years employed this method in their investigations on Mo and Cu deficiencies[10].

Following the establishment in the 19th century of the general principles of photosynthesis in the life cycle of green plants, scientists shifted to a systematic determination of the complete inorganic elements essential to their growth. At first attention was centered on the nutrient solution, since impurities introduced by the salts and other contaminants could vitiate the results. Initial studies by plant physiologists seemed to indicate that, generally, the mineral needs of the higher

[8] Sachs, J., *Landwirtsch. Versuchs-Stat.*, **2**, 219; **3**, 30–44 (1860).
[9] Knop, W., *Landwirtsch. Versuchs-Stat.*, **2**, 65–99, 270–93 (1860).
[10] "The water-culture method," *Circ. Univ. Calif. Coll. Agr.*, No. 347, 1938.

plants in the tests were satisfied by seven inorganic elements, that were provided by the nutrient solution plus carbon, hydrogen, and oxygen derived from the air and water. Not until subsequent research was it suspected that impurities in the salts or from other sources supplied the additional, needed nutrient elements. The seven inorganic elements originally regarded as sufficient by Knop and Sachs were present in the form of salts in the nutrient solution. Until 1920 it was believed that the total nutrient requirements were fully satisfied by ten essential elements: the seven inorganic elements supplied by the cultural solution as salts plus carbon from carbon dioxide and hydrogen and oxygen from water. Our current, definite knowledge that plants require at least seven other elements in trace amounts is a comparatively recent development of tremendous scientific interest and practical application.

2

BIOPHYSICOCHEMICAL
RELATIONSHIPS

Atoms; Molecules; Elements

Since the physical sciences are extensively employed in experimental work on plants, a knowledge of some pertinent fundamental facts of physics and chemistry is helpful to an understanding of at least the elements of physiological processes. Nuclear physics has been a great aid to the understanding of atomic structure and properties and is the basis on which currently sound explanations have been developed. The student who desires to delve deeply into plant structure and function is referred to textbooks on plant physiology. Here we can only give surface consideration to the subject. In order to acquire a comprehensive knowledge of the composition and nature of protoplasm, cell walls, tissues, and organs, it is very helpful to study both the large and the microscopic structures of plants and the processes involved in the formation, modification, or destruction of these structures; such knowledge requires an understanding of the nature and properties of the invisible chemical and physical units which constitute all matter. A brief consideration of such units follows.

Plants as well as all other things in the world consist of substances with which the science of chemistry is concerned. To properly understand the mechanism of life requires some understanding of the professional terminology relating to the constitution of matter. Earth, air, fire, and water were at one time in the past called "the four elements of the material world." Chemists now apply the term "element" to all the simple chemically indivisible substances which make up matter. The number of chemical elements listed in the periodic table is 92 plus a few radioactive elements which spontaneously produce matter unlike themselves. Some of the known elements exist naturally in the form of gas, as for example, nitrogen, oxygen, and hydrogen; some, as metals, such as gold, copper, silver; and others occur only in combination with other elements.

Chemists have been investigating and analyzing materials for over a century and have learned that most natural substances consist of at most 25 commonly occurring elements. The following list consists of these most important chemical elements. Most of them are identified with Latin names. Chemists also use symbols in place of names, as shown in the table, the symbol being frequently the first letter or two of the Latin name (Kalium is an Arab name).

Every chemical element consists of particles known as atoms which are millions of times smaller than the smallest fragments visible under the highest magnifying power of the microscope. Atoms of all elements have different weights. The physicist declares that even an atom is divisible, that it is an infinitely minute microcosm with a definite structure: a nucleus in its center charged with positive

TABLE 2.1. SOME IMPORTANT CHEMICAL ELEMENTS

NAME	SYMBOL	NAME	SYMBOL
Aluminum	Al	Manganese	Mn
Boron	B	Molybdenum	Mo
Calcium	Ca	Nickel	Ni
Carbon	C	Nitrogen	N
Chlorine	Cl	Oxygen	O
Cobalt	Co	Phosphorus	P
Copper (Cuprum)	Cu	Potassium (Kalium)	K
Fluorine	F	Selenium	Se
Hydrogen	H	Silicon	Si
Iodine	I	Sodium (Natron)	Na
Iron (Ferrum)	Fe	Titanium	Ti
Magnesium	Mg	Vanadium	V
		Zinc	Zn

electricity (a proton), and one or more negatively charged units of electricity (electrons) in orbit around the nucleus.

The term molecule is used to describe the smallest particles capable of existing as separate entities not necessarily combined with other molecules. Each molecule consists of atoms which are specific in kind and structural arrangement. The properties of atoms are likewise determined by still smaller units called electrons, protons, and neutrons.

A chemical compound refers to a combination of at least two elements held together by chemical affinity or electrical or magnetic forces and in a definite proportion. An example of a simple compound is table salt or sodium chloride which consists of one atom of sodium (Na) closely united by electrical or magnetic forces with one atom of chlorine (Cl). The molecule of salt is represented in chemical shorthand by the symbol NaCl.

Essentiality of Nutrient Defined

The word "essential" as applied to trace elements needs an explanation. It is used to describe the absolute nutritional indispensability of a chemical element. Intensive, exceedingly refined analytical methods, chemical and instrumental, were employed by modern research workers to assess the requirements of plants and animals for certain nutrient elements. For this purpose the investigators had to exercise skill and extreme precautions to avoid contaminated chemicals and laboratory equipment, and use the most painstaking refinement of plant-culture technique because they were dealing with infinitely small amounts of the element present in the sample. For example, nutrient salts had to be especially purified, and great care used in preparing water by distillation. Then, to prove essentiality of the micronutrient under study, it was necessary to demonstrate the following attributes established by D. I. Arnon: (1) that the plant under study could not complete its life cycle without the element in question; (2) that the element in question was directly required for the nutrition of the plant; and (3) that no other chemical element could substitute for it.

Besides gases, solids, and liquids, the other states of matter of particular importance comprise solutions, emulsions, and colloidal suspensions. Man has been able to use these to reproduce, artificially, numerous physiological processes and thereby discover characteristic phenomena of living organisms.

Solutions

If the molecules of salts, acids, and bases are dissolved, they dissociate into units called ions, which are electrically charged atoms or groups of atoms called radicals. An ion carrying one or more positive charge is called a cation, for example, $(NH_4)^+$, Na^+, Ca^{2+}. An ion with one or more negative charges is called an anion, examples of which are $(SO_4)^{2-}$, Cl^-, $(PO_4)^{3-}$, $(OH)^-$. When two kinds of atoms remain in combination throughout a chemical reaction and behave as if they are a single ion, that group is called a radical, for example, $(SO_4)^{2-}$, $(NO_3)^-$, $(NH_4)^+$. A solution is defined as a system in which one component (the dissolved chemical) called the solute, is distributed uniformly throughout a second component (the solvent) in the form of molecules only or molecules and ions. For example, if potassium chloride, KCl, is dissolved in water, some of its components become divided and separated in the form of molecules, KCl, and ions, K^+, Cl^-. Thus, particles in true solutions are molecules or ions or both, and since they are usually less than 0.001 of a micron (μ) in diameter ($1\mu = 0.001$ millimeter), they are far beyond the range of ultramicroscopic visibility.

Solutions make up important media in living plants, such as those in which the major movement of materials occurs from cell to cell. Plant solutions may contain acids, bases, salts, and soluble organic compounds and are of major importance to plants. Soil water speeds up the disintegration of minerals into ions. Although soil water does not carry the mineral substances into the roots, it does serve as a solvent in which the minerals dissociate into molecules and ions, and as ions, the minerals are able to enter or leave the roots independently of the water.

Six kinds of solutions are known: gas in gas; gas in solids; solid in solid; gas in liquid; liquid in liquid; and solid in liquid. The three last are of greatest importance in plants.

Ions; Emulsions; Colloidal Suspensions

Other forms of matter encountered in the study of plants are emulsions and colloidal suspensions. If clay is placed in an adequate amount of water and shaken vigorously, a mixture named a colloidal suspension results. This is intermediate between a suspension and a solution. A colloidal suspension is a system in which the particles of the dispersed phase (clay) are entirely separated by the particles of the dispersion medium (water). This type of matter is very important in the life processes of plants.

In emulsions the particles are also molecular aggregates, as in a colloidal suspension, and are larger than 0.1 micron in diameter. Emulsions are not stable unless an emulsifying agent is in the system. Milk is a good example of an emulsion.

3

SOIL-PLANT
RELATIONSHIPS

Anions; Cations

Most plant physiologists maintain that soil minerals enter the roots as individual anions or cations. Researchers using radioisotope P^{32} have shown that while most of the P^{32} ions placed in the soil entered the root hairs and cells in the region of cell elongation, only a small amount was absorbed by the meristem cells (Meristems are tissues in which the cells can divide indefinitely and are located at growing points such as the tips of stems and roots).

The current concept of how ions move into and inside a rooted green plant is that a transpiring plant activates a pull-passive mechanism, which causes water to move into the root absorption cells. Thereby, both water and ions move together in a mass flow into the xylem and upward to the rapidly growing buds, fruits, and seeds.

Many ions move into root cells against a gradient and not by diffusion, as formerly believed. Large accumulations of ions in root cells are found in the meristematic region of the root tip. This is called primary or active absorption. A secondary absorption occurs when H^+ ions in the cells are exchanged for cations, such as K^+, Ca^{2+}, Mg^{2+}, etc. in the external solution; while anions such as $(OH)^-$ and/or

HCO_3^- in the cells are exchanged for anions such as $(SO_4)^{2-}$, Cl^-, $(NO_3)^-$ in the external solution.

Green Leaves (Chlorophyll)

Green leaves are a familiar feature of plant growth; they are the organs of photosynthesis by which the plant absorbs the energy in sunlight and converts carbon dioxide and water into sugars and starch food reserves. The green color is caused by highly specialized, energy converters called chlorophyll pigments. Chlorophyll was isolated and named from the Greek words for "green leaves" by J. Pelletier (1788–1842) and J. B. Caventon (1795–1877) in 1817, although its importance was at first not appreciated. J. R. Mayer, who propounded the law of the conservation of energy, clearly saw the relation of this energy concept to the photosynthetic process. In 1845 Mayer pointed out that plants fixed the energy of sunlight to later supply this energy as the source on which mankind depends. In their effectiveness and efficiency leaves surpass man's most elaborate chemical process factories.

The first important generalization in the 19th century resulted from the work on photosynthesis, and it led to the unravelling of some of the complex relationships between plants and animals. Priestley and Ingenhousz found that green plants gave off oxygen in sunlight. Senebier[1] showed further that the plants absorbed carbon dioxide during this process. Theodore de Saussure in his classic "Recherches chimiques sur la vegetation" (Paris, 1804) demonstrated by quantitative measurements that the carbon in the dry matter of plants came almost entirely from CO_2, and, equally important, that the remainder of the dry matter, except the soil minerals, came from water.

Chlorophyll contains magnesium, an element which is essential to its formation and existence, and although equally essential iron is not a component of chlorophyll.

Roots[2]; How Plants Feed

The root system is normally embedded in a conglomerate of soil colloids which are in contact with one another. They can exchange

---

[1] Senebier, J., "Recherches sur l'Action de la Lumiere Solaire dans la Vegetation," Geneva; Paris, 1788.
[2] Schweigert, H. A., "Mechanism of the Reaction of Trace Elements," Rev. 25, II, 1956.

cations and anions among themselves. Thus linked, they could be considered as active, contact colloid chains. Barring any outside force, all the soil colloids of the chain would, in a short time, be in equilibrium with respect to their content of ions and water. But around every soil colloid in contact with root hairs, an environment develops which is continually changing, apart from outside influences such as temperature, pressure, moisture, buffering in the medium, variations in concentration, and other influences. Accordingly, ions are released by the colloid while others are being reabsorbed, perhaps by exchange from the root cells or from the surrounding soil solution. Now, if the soil colloids are linked with one another and arranged as in a mosaic, and if any one of the links of the chain is altered, it follows that the linkages in all other parts of the chain will be broken, as well as the state of equilibrium. Under normal conditions this can hardly happen during the course of a growth cycle. It must be permanently destroyed for such to occur.

Another modern viewpoint on mineral absorption through the roots is that many of the absorbed ions remain in the uncombined state in the vacuolar sap of the cells, particularly the cations K^+, Na^+, Ca^{2+}, in concentrations many hundreds of times those in the soil solution. Metabolic activity controls the amount of ions accumulating in the vacuoles and holds them against a gradient. This effect seems to be associated in some manner with the process of respiration.

Active absorption of ions, it is theorized, occurs by means of a carrier (an unknown organic constituent), which diffuses with its ions into the vacuolar sap where it releases them. The vacuolar membrane is permeable to the carrier but not to isolated ions. Thus, the vacuolar sap ions are absorbed irreversibly.

4

TRACE ELEMENTS IN NUTRITION

Historical Background

Life and perhaps civilization are closely allied with photosynthesis, but the popular concept puts stronger emphasis upon the requirement of plants for certain mineral elements derived from soil. The probable explanation for this is that by fertilization and soil management practices, it is possible to modify the soil medium in which plants grow and to increase the yield of plants through an increase in the supply of mineral fertilizer elements. The fertilizer industry refers to such elements as "plant food." Actually, however, there is no difference in the mineral nutrients of plants and that of animals—both require carbohydrates, fats, proteins, and minerals. The fundamental difference between plants and animals in this regard is that the plant is autotrophic, that is, it is able to synthesize its own food as well as that required by animals. The inorganic salts indispensable for plant growth serve partly as constituents of the synthesized organic substances, and partly as catalytic agents of the chemical reactions involved. They may also be involved as essential components of the metabolic processes concerned with growth and synthesis.

As previously mentioned, the scientific basis of plant nutrition is a comparatively recent development. The water-culture method of

Sachs and Knop has been in time improved, refined, and used by a number of eminent plant physiologists. In the early stages of the water-culture method the impurities in the nutrient salts and other sources of contamination were not completely realized. The first indication that the plant needed trace amounts of chemical elements not included by Sachs and Knop in their classical water-culture research was discovered by Miss K. Warington at the Rothamsted Experimental Station in Harpenden, England. She was growing broad bean plants in different nutrient solutions of the same composition, but certain solutions contained a minute amount of boron besides the traditional seven inorganic elements then considered adequate for normal growth. A striking difference in growth and vigor was apparent in the plants which received the boron. Dr. D. R. Hoagland later confirmed the result through similar tests conducted in his laboratory at the University of California. Some years later Dr. Hoagland, while engaged in a study of a serious disease of fruit trees often called "little leaf," discovered partly by accident that the condition was caused by a deficiency of zinc, and a similar deficiency was found to be the cause of "mottle leaf" of citrus trees. Soon zinc deficiencies were also observed by workers in other areas and on other species of trees and field crops. The essentiality of zinc and boron was definitely established in due time. Before this happened, several other investigators had established evidence that manganese was an essential nutrient element for plants, and, shortly thereafter, copper was added to this list. P. R. Stout[1] of the University of California published results of an investigation, which were confirmed by Australian scientists, that certain higher plants, if not all, when grown in nutrient solutions or in soil, responded remarkably to very small applications of molybdenum. These investigations and their results have become of great practical significance in agriculture in terms of substantial increases in crop production in deficient soils, even though such deficiencies are present only under special soil conditions and only in certain soils (see Table 4.1).

Rooted green plants growing in soil tend to absorb all the mineral elements present in the soil solution. The soil contains varying amounts, sometimes in parts per million, of far more minerals than the sixteen at present recognized as being essential to all plants if they are to complete their life cycle. Chemical analysis of plants has revealed some sixty elements being present in their tissues. Since no definite role is

[1] Stout, P. R. and Johnson, C. M., *Soil Sci.*, **81**, 183–190 (1956).

TABLE 4.1. DISCOVERY OF ESSENTIAL MICRONUTRIENTS IN PLANT NUTRITION

MICRONUTRIENT	YEAR DISCOVERED	FIRST REPORTED BY
(1) Manganese	1922	J. S. McHargue
(2) Boron	1923	K. Warington
(3) Zinc	1926	A. L. Sommer and C. B. Lipman
(4) Copper	1932	C. B. Lipman and G. McKinney
(5) Molybdenum	1939	D. I. Arnon and P. R. Stout
(6) Chlorine	1954	T. C. Broyer, A. B. Carlton, C. M. Johnson, and P. R. Stout
(7) Sodium	1957	P. F. Brownell and J. G. Wood

Literature references to the first publication of each discovery are as follows:

(1) McHargue, J. S., "Role of manganese in plants," *J. Am. Chem. Soc.*, **44**, 1592–1598 (1922).

(2) Warington, K., "Effect of boric acid and borax on the broad bean and certain other plants," *Ann. Botany (London)*, **37**, 629–672 (1923).

(3) Sommer, A. L., and Lipman, C. B., "Evidence of the indispensable nature of zinc and boron for higher green plants," *Plant Physiol.*, **1**, 231–249 (1926).

(4) Lipman, C. B., and McKinney, G., "Proof of the essential nature of copper deficiency," *J. Pomology*, **10**, 130–146 (1932).

(5) Arnon, D. I., and Stout, P. R., "Molybdenum as an essential element for higher plants," *Plant Physiol.*, **14**, 599–602 (1939).

(6) Broyer, T. C., Carlton, A. B., Johnson, L. M., and Stout, P. R., Chlorine—a Micronutrient Element for Higher Plants," *Plant Physiol.*, **29**, 526–532 (1954).

(7) Brownell, P. F., and Wood, J. G., *Nature*, **179**, 635 (1957).

as yet known for these other mineral elements in the life cycle, they are considered unnecessary, and their presence is explained as due to a lack of selectivity; "plants do not have a soul to think with" is the way Aristotle put it.

Up to 1920 plant physiologists believed that only the ten elements first reported by Sachs and Knop were essential to normal plant life, namely, C, H, O, N, P, K, S, Ca, Mg, and Fe. Then, in the 1920's,

1930's, and 1950's, very precise analytical procedures disclosed that the following additional elements are needed in trace amounts: boron (B), zinc (Zn), copper (Cu), manganese (Mn), molybdenum (Mo), chlorine (Cl), and sodium (Na). Table 4.1 lists the elements, the year discovered, and the investigators who first reported on each.

During the past 40 years trace elements or micronutrients have become increasingly prominent as elements indispensable to the health of plants, animals, and man, and to agricultural productivity. They help maintain the productivity of land already under cultivation, and, as fertilizer components, they help restore fertility to exhausted, infertile soil and enable new tracts of land to be developed for crop production. Zinc, copper, and molybdenum have made it possible to utilize vast unproductive areas, particularly in Australia, for agricultural purposes. The results have been spectacular. In many sections of the western region of the United States, zinc is now second in nutritional importance only to nitrogen as a fertilizer material, although, of course, not in total tonnage consumed.

Biochemical investigations in many countries continue to reveal vitally important secrets regarding the relationship of trace elements to life processes. Recent examples are the relation of cobalt to the nitrogen-fixing capabilities of the leguminosae, and that of selenium to the prevention of white muscle disease in mammals. To judge from the many enexpected discoveries of the role of certain trace elements in the life processes, it may not be surprising if future research discovers, among the 60 or more chemical elements revealed by analysis as being present in the cells of plants and animals, that some have a significant, essential function in the complex processes of living organisms. The element chromium appears to be a candidate for that elevation, and further research may show that vanadium, fluorine, silicon, and aluminum may in time also qualify.

Dr. W. D. McElroy[2] has emphasized that past investigators in their efforts to identify and prove the essentiality of nutrients, for what is referred to as "normal" growth and development, assumed that all the units in a population of plants or animals were identical and hence had the same nutritional requirements. Current evidence shows that large, individual variations exist in nutritional needs. Besides, the quantitative studies on different tissues indicate that small changes in the diet may often lead to unexpected, large effects on enzyme systems.

[2] McElroy, W. D., Lecture, Nat. Res. Council, Vol. 73, No. 8, 1958.

Nutritional individuality must henceforth be considered an important factor in health and disease. The fact that trace elements are required in small amounts indicates that they function in some catalytic role, usually as part of an enzyme system, somewhat similarly to that of the organic micronutrients called vitamins. Research has shown that most of the trace elements do have an enzymatic role, iodine and cobalt perhaps being exceptions; iodine is a component of the thyroglobulin molecule and cobalt is a part of vitamin B_{12}. Everyone, it seems, knows that vitamins are important to health, but the vital role of trace elements is less well-known.

Good health and vigor reflect a positive state of well-being and not merely the absence of disease. To attain this active state requires perfect functioning and teamwork of the many organs and tissues in a plant or animal. This, in turn, depends upon an adequate ingestion of a large variety of substances. A living organism is delicately balanced and can easily be upset by unbelievably small nutrient deficiencies. Unknown until quite recently, the vital role trace elements play has now been definitely established.

Perhaps it may eventually be proved that some of the chronic and fatal diseases of humans and livestock are the result of accumulations, deficiencies, or displacements of specific trace elements. A number of authorities have thus hypothesized about the possibilities. One of these is Dr. Henry A. Schroeder, an experimental physician known for his studies on man and his environment. He reports[3] that by means of spectrographic analyses 29 chemical elements have been identified as forming the basis of modern thinking on this subject. Some are abnormally present in the tissues of man, some accumulate with age, some have organ specifications, and some are toxic. He believes five trace elements are essential to the life and health of man. They are manganese, iron, cobalt, copper, and zinc, and a sixth, chromium, seems to have good qualifications to belong to this group. These metals are chelated to protein or other organic chemicals in the body. Three other trace elements, titanium (Ti), nickel (Ni), and vanadium (V), act as catalysts in biological reactions but as yet are not recognized as essential for life. All these nine trace metals are relatively nontoxic, ubiquitous, reactive, and nonaccumulative to man. Nickel is found in all plants and plant products but not in meat, milk, or eggs. In man it appears quite frequently in the skin and intestine and does not

[3] "The Spex Speaker," Vol. 6, No. 2, 1961.

accumulate with age. Cobalt is part of the molecule, vitamin B_{12}, and has been proved essential to ruminants, which require a few micrograms of it daily to supply the microorganisms in the rumen that produce the vitamin B_{12}. Many other metalloenzymes have been discovered by recent research to be important in human and animal nutrition. Several peptidases depend upon manganese to help in their digestion of protein. Copper oxidases are well known to nutritionists. Six zinc metalloenzymes and two molybdenum enzymes have been identified as being required by mammals.

The Trace Element Problem

Investigations of trace element problems have most often been of two purposes: (1) to explain crop failures, and (2) to determine the effects on plant growth of elements other than those already recognized as essential. As a consequence of (1), much practical data has been collected regarding the effects of trace elements on crops, but relatively little information has been found pertaining to their role in the intricacies of plant growth. From (2), only slow progress has occurred, because of the difficult nature of the problem. Another involvement needing elucidation is the role of the trace elements as catalysts in the enzyme systems of plants and animals. Many specialists in biochemistry now tend to believe that the chief role of trace elements is as catalysts in the biochemical processes governing growth and reproduction.

Some soils are naturally deficient in one or more trace elements, although no soil is absolutely exempt. In the United States the soils in the area around the Great Lakes is notoriously deficient in iodine, which is responsible for the relatively large incidence of goiter which prevails locally. Another area known to have deficiencies is the coastal plain soils along the Atlantic Ocean and the Gulf of Mexico.

In the literature on deficiencies in this country the student will find maps of the United States, which show that in 30 states a deficiency of manganese was reported and a note indicating that the total deficient area comprises 13 million acres; that boron was reported deficient in 44 states with an area of 12 million acres involved. Such reporting exaggerates the involvement and tends to excite farmers to apply trace elements indiscriminately to their fields. No such pinpointing of deficiencies over millions of acres of soil is really justified on present information.

Trace Elements and Soils

The ultimate source of trace elements is the soil. It may be poor or deficient in them as a result of erosion, leaching, removal by crops, or the absence of mother rocks containing them. Early civilizations began in river valleys or on deltas at the mouths of rivers. The soils of such areas are usually rich in many inorganic elements, which flooding rivers carry in the silt, and clay washed down from a broad and diversified watershed. As populations expanded, they had to seek new sites for settlement. These were beyond the first choice settlements, and the soils were much less fertile and relatively poor in nutrient elements. Crops raised in reclaimed land show deficiency diseases, as attested by experience on such land in Holland, Denmark, and Australia, so much so that the term "reclamation disease" was coined to describe them.

It is possible to have essential elements in a soil without any assurance that plants are absorbing them in sufficient quantities. Many soil factors will influence the uptake: pH or degree of acidity or alkalinity of the soil, organic matter content, moisture, pore space, proportion of clay and silt, size and type of soil microbial life. All these factors will have some effect on the availability of the trace elements to plants. Iron is usually combined with phosphate in an unusable form if the soil is acidic. Liming can modify the acidity and cause the iron-phosphate combination to release the phosphate radical and make the iron available. Zinc combines with soluble phosphate to form insoluble zinc phosphate. The availability of boron is enhanced if the soil reaction is on the acid side, and that of molybdenum is enhanced by adding lime to neutralize the acidity. In some peat soils copper is deficient. Manganese deficiency prevails in certain peats and in soils having a high content of organic matter.

To correct a deficiency of trace elements is not a simple matter, particularly because of the very small amounts involved. The addition, if excessive, may easily prove fatal to the plants. For example, experience with the potato crop has shown that on some soils as little as 20 pounds of borax per acre was ruinous. No general rule can be given for supplying the proper amount of trace elements, because every species of plant has its own needs and tolerances, and the effects of the various trace elements are interrelated. For example, copper and cobalt work in association with iron in the production of the hemoglobin of red blood cells. An excess of molybdenum in an animal's diet may cause severe copper deficiency symptoms, even though the supply of copper is sufficient.

The quantities of trace elements in ordinary soil amendments[4] are seldom sufficient to affect the total content of a soil but may at times appreciably influence the soluble or available fraction. Lime $(CaCO_3)$ is the amendment applied in greatest amounts to soils in the podzolic zone; this is done usually every 5–10 years at the rate of about two parts in 1,000 of the cultivated, plough depth layer. Limestones commonly contain about 1,000 ppm strontium (Sr); dolomites, generally less, sometimes much less than 100 ppm; Mn occurs in amounts of between 100 and 1,000 ppm; B seldom exceeds 10 ppm; and Co and Ni do not exceed 5 ppm. If quartz is present, it is accompanied by some zircon (Zr). An amount of 10 ppm in a limestone means an addition of 0.02 ppm to the soil. These levels are negligible when compared with total soil content.

A Czech investigator, S. Nikolic[5], believes the list of essential elements for plants should be markedly increased beyond the 18 currently accepted. He carried out a series of pot experiments over a 3-year period on certain trace elements including nickel, cobalt, and fluorine, using mixtures of NPK incorporating the following compounds: nickel chloride, cobalt chloride, sodium fluoride, and ammonium fluoride. He used two soil types, a calcareous, black earth (chernozem), and an acid or podzol soil. The test crop was the oat.

On both soil types the generally recognized NPK fertilizer gave a consistent increase in yield (something like a 6-fold increase), but an even greater increase in yield of grain and straw (8-fold and more) was obtained from the NPK mixture incorporating one or other of the trace elements. Hence, he concluded that the oat plant makes better use of the major nutrients (N, P, K) in the presence of these trace elements.

Nikolic raised an interesting point regarding nomenclature of nutrient elements: what exactly is to be understood, he asks, by such terms as oligo-elements, mikronährstoffe, elements-traces, spüren-elemente, trace elements, micro-elements? These several terms are apparently used in different countries to designate the group of nutrient elements generally recognized as essential even though needed in extremely small amounts.

[4] Mitchell, R. L., Loc. cit., 1955.
[5] *Sixth Intern. Congr. Soil Sci.*, Vol. **D**, Commissions IV and VI, p. 85, Paris (1956).

He regards all these terms as not being identical and equally applicable in all circumstances, since the various elements are not all comparable in their effects on plants and animals. R. L. Mitchell has stated that the term "trace elements" may refer to the biological significance of such elements to plants and animals, or to the absolute content of the soil. No definition is satisfactory from all viewpoints[6].

Dr. D. I. Arnon has specified the criteria by which the essentiality of a micronutrient is to be evaluated. According to his specifications the term "micronutrient" may be restricted to a number of trace elements, analogous with the nine major and secondary elements collectively called "macronutrients." "Trace elements" themselves then make up a broader and more general group within which the micronutrients represent something more restricted and more tangible. Classifications in this sense have already appeared in the U.S.S.R., Germany, the United Kingdom, and other countries.

Mineral Nutrition of Plants

Plants are living organisms. Like animals and humans, plants ingest foodstuffs and incorporate them into their organs and tissues and convert some into energy. Plants can be healthy or sickly, suffer from acute diseases or minor ills, or thrive on a balanced set of nutritive elements. Modern science has shown that plants and animals require almost the same nutrient elements (see Table 4.2. The chemical composition of 100 bu. corn). When the diet of humans or livestock lacks one or more essential nutritive elements for a significant length of time, a serious nutritional disease will develop. For example, rickets in animals is caused by a deficiency of calcium, phosphorus, and vitamin D; nutritional anemia results from a lack of iron. Plants show a nutritional deficiency by definitive symptoms that occur particularly in their leaves, and these have been studied, identified, and catalogued by plant scientists. For example, a lack of nutrient iron shows up as chlorosis. It is now recognized that plants require a variety of essential mineral nutrients in varying amounts.

[6] Mitchell, R. L., "Chemistry of the Soil," ACS Monograph, No. 126, Reinhold Corp., N.Y., p. 253, 1955.

TABLE 4.2. AVERAGE AMOUNT OF SPECIFIED ELEMENTS IN COMPOSITION OF 100 BU CORN (MAIZE)

Yield

4,000 lb stover 1,400 lb cobs 5,600 lb shelled corn 5,200 lb roots and stubble	16,200 lb dry matter and if all the water is excluded, this figure drops to 14,200 lb dry matter in the crop growth.

SUBSTANCE	SYMBOL	LB/ACRE	APPROX. EQUIVALENT
Furnished by Air and Water			
Water	H_2O	4.3–5.5 million	19–24 in. of rain
Oxygen	O_2	6,800	Air is 20% O_2
Carbon	C	5	Amount C contained in 4 tons of coal
Major Nutrients from Soil and Fertilizers			
Nitrogen	N	130	4 bags (100 lb) of a 32% nitrogen fertilizer
Phosphorus	P	22	250 lb of 20% superphosphate
Potassium	K	110	200 lb KCl (muriate of potash)
Sulfur	S	22	22 lb brimstone sulfur
Magnesium	Mg	33	330 lb Epsom salts
Calcium	Ca	37	93 lb limestone
Trace Elements from Soil and Fertilizers			
Iron	Fe	2	2 lb nails
Manganese	Mn	0.3	1 lb potassium permanganate
Boron	B	0.06	¼ lb common borax
Chlorine	Cl	Trace	Enough in rainfall
Iodine	I	Trace	1 oz tincture of iodine
Zinc	Zn	Trace	Shell of 1 dry cell battery
Copper	Cu	Trace	25 ft of #9 copper wire
Molybdenum	Mo	Trace	1 oz. sodium molybdate

Source: F. E. Bear and Textbooks.

Several investigators knew and recorded this relationship long before the modern era. For example, Carl Sprengel[7], a German scientist who preceded J. von Liebig, understood remarkably well for that period the elements of agronomy and the mineral nutrition of plants. Long before Liebig, who receives universal credit for the concept of mineral nutrition of crops, Sprengel published views on mineral nutrition, most of which are in line with modern concepts. His thoughts on soil-plant relationships, as expressed in his "Bodenkunde[8]" reveal this:

"The different parts of plants contain their elementary constituents in widely varying proportions. The grain of wheat is much richer in phosphorus, sulfur, nitrogen, calcium, potassium, sodium, and chlorine, while wheat straw contains only small amounts of these elements but much more of silicon. It follows, therefore, that if we wish to grow wheat with a high yield of grain, the soil, or its fertilizer, must be amply provided with the seven elements previously mentioned. Experience demonstrates that this is always actually the case."

Sprengel emphasized that plants differ greatly in their preferences and needs for the various nutritive elements of the soil. Again, to quote him:

"If, however, an element necessary for the chemical constitution of this or that plant is completely lacking in the soil or fertilizer, it is impossible for it to grow, for, so far as we know, no necessary element in the processes of vegetation can be replaced by another, or still less be generated by transmutation . . . . It has recently been asserted that the bases, lime, magnesia, potash and soda, can replace one another . . . I must, however, contradict this assertion." (From his, Lehre vom Dünger (1845), pp. 58–59.)

The chemical analysis of a plant can, at best, reveal its chemical constituents, but it does not necessarily indicate the particular quantitative requirements (see Table 4.2). Root hairs seem to indiscriminately absorb the ions present in the rhizosphere; similarly, nothing seems to prevent gases in the atmosphere from entering the plant proportionate to their solubilities in the plant sap. The analysis of the plant may provide a clue to the nutritional needs and may permit an estimation of the amounts of mineral nutrients removed from the soil at each harvest. The data in Table 4.2 are an example of this possibility.

[7] "Bodenkunde in Chronica Botanica," Vol. VIII, p. 235, 1944.
[8] Sprengel, C., "Lehre vom Dünger," Leipzig, 1845.

The chemical composition of a plant is the result of an interaction between its genetic inheritance and the environment in which it grows and completes its life cycle. In some plant species, for example the sugar beet, the sugar content is determined chiefly by genetic factors and only slightly by environmental factors. In other crops such as rye grass, for example, variations in its protein content can be caused mainly by nutritional factors.

Genetics and Crop Yields

Scientists are recognizing that corn hybrids differ in their response to specific levels of nutrients in the soil; certain hybrids have greater ability to extract trace elements from the soil and translocate them to different tissues of the plant. This is quite the situation when the soil is considered to contain adequate amounts of secondary and trace elements.

Research at the Pennsylvania State College[9] shows that accumulations of plant nutrients by corn hybrids depends upon independent mechanisms controlled by genetic factors. The investigators employed varying quantities of calcium and strontium and other nutrients in their experiments with several genetic lines of field corn.

The studies definitely showed that hybrids which accumulate relatively large or small concentrations of one element do not accumulate all elements in the same relative manner. It was then deduced that the differences in accumulation of each element result from independent mechanisms governed by varying genetic factors.

Some data from these experiments indicate that even when different amounts of a particular nutrient are made available to a plant during its growth, specific hybrids or genotypes retain the same ability to accumulate high or low amounts. For instance, a hybrid with a low accumulation characteristic might use less phosphorus to produce a crop than a high accumulating hybrid.

The varied response of corn hybrids to essential nutrients shows up when varieties are grown side by side on a soil supplying almost marginal levels of a secondary or a trace element (as in the case on soils inherently low in such nutrient element). Other similar situations

[9] *Science for the Farmer*, **8** (Spring, 1966).

are: (1) a variation in the soil pH or a nutrient imbalance; (2) a temporary shortage resulting from increased nutrient demand on the soil by a high yielding hybrid over a relatively short time. In such situations a hybrid variety that thrives under optimal conditions may respond far below its potential capacity; or, a variety with a relatively low potential may show a better yield than the first variety on soils of low fertility because it has a more efficient absorptive system.

The lack of quick, routine, and reliable tests to determine secondary and trace elements handicaps growers who otherwise would have few problems in providing optimal growing conditions for corn hybrids. The same applies to hybrids of other species. Up till now such tests have not been developed. Because of this lack, it is possible to infer the following: (1) in many areas of the United States, and undoubtedly also in other countries, trace element soil problems have not been precisely defined, and this is particularly true where crop production goals are being yearly and rapidly increased. (2) Agronomists and farm advisers, when making fertilizer recommendations, apparently often fail to consider the differences in yield potential among hybrid varieties. The consequence has been a decision that no trace elements were needed, when as a fact they were already below the optimum content for the variety involved. (3) Farmers feel that owing to the emphasis currently placed on the importance of trace elements in crop production, they cannot afford to take the risk of not having an optimal amount, so they apply them indiscriminately. This practice can in turn create problems. For example, the continued application of boron to a soil can cause a buildup to toxic levels; similarly, applications of manganese to a highly acid soil can likewise be harmful.

For the purpose of discussion it is possible to visualize an area in between the perfect environment favorable to the harmonious development of a crop to its level of maximal yield and that completely unsuitable environment in which a crop cannot possibly attain its full growth and maximal yield. Between these extremes a crop may develop to maturity and bear its harvested yield, but the plants will have a variable composition, that is, the organic and inorganic elements in its various parts will vary from the norm established under ideal conditions. In practice, satisfactory crop yields can be produced under different conditions, and this is why it is difficult to refer to an "optimal composition" of the yield. At best, yield level may be associated with a more or less distinctly outlined, optimal level of chemical composition which is then considered "the norm." This is further explained by the following three situations.

An application of chemical fertilizer may influence the crop in three ways:

(1) An influence on the environment of the rhizosphere may lead to changes in the interaction between the plant and its environment. Such changes are generally chemical but may very well involve physical and biological factors. An application of lime or organic manure can definitely alter the conditions in the root zone. However, in principle, all fertilizers have a similar effect, more or less, on the conditions in the rhizosphere.

(2) When nutrient elements are absorbed into the plant, they are utilized for the elaboration of organic materials and diverse plant tissues. These in turn affect the subsequent absorption of ions by the plant and eventually determine the amount of water taken up and distributed within the plant. The water content affects the concentration of the nutritive elements in the plant, thereby affecting the chemical composition of the plant.

(3) The uptake of inorganic ions is associated with the quantitative and, to some extent, qualitative synthesis of organic matter. This explains why differences in the organic constituents of the plant may be caused by varying the amount of inorganic salts applied to the soil.

Plants require at least 15 *mineral* elements for normal growth and development. Those needed in relatively large amounts are named macronutrients (N, P, K, Ca, Mg, S); others needed in very small amounts occur mainly in parts per million and are called trace elements or micronutrients. The trace elements considered essential to plant life are: boron (B), copper (Cu), iron (Fe), chlorine (Cl), manganese (Mn), molybdenum (Mo), sodium (Na), and zinc (Zn). Animals obtain most of their micronutrients from plants. If to the number of trace elements required for nutrition by higher plants are added those deemed necessary for some species of microorganisms, such as bacteria, fungi, algae, and those for animals, the list will include eight more trace elements, namely, cobalt, iodine, selenium, fluorine, silicon, vanadium, aluminum, and chromium.

The average amount of a trace element varies considerably in different organisms. For example, iron or manganese in dried plant material may be present at about 100 ppm, whereas the amount of zinc or copper may average only 10 ppm, and that of molybdenum only 1 or 2 ppm or less (see Table 4.3). The list of nutrients is not to

TABLE 4.3. RANGE OF CONTENTS OF TRACE ELEMENTS IN HERBAGE PLANTS

ELEMENT	% OR PPM DRY WT	HERBAGE
Na	0.002% 2.12%	mixed pasture Italian rye grass
Cl	0.015% 2.05%	mixed pasture perennial ryegrass
Fe	21 ppm 1,000 ppm	ryegrass alfalfa
Mn	9 ppm 2,400 ppm	meadow grass alfalfa
Zn	1 ppm 112 ppm	grass alfalfa
Cu	1.1 ppm 29.0 ppm	mixed pasture red clover
Co	0.016 ppm 4.7 ppm	subterranean clover subterranean clover
I	0.069 ppm 5.0 ppm	alfalfa pasture
Se	0.01 ppm 4,000.0 ppm	grasses, clover mixture weed species
B	1.0 ppm 94.0 ppm	grasses alfalfa
Mo	0.01 ppm 156.0 ppm	red clover clover

Sources: Numerous investigators in published data.

be considered final, but presumably, if others are added, they will be required in tiny amounts, and exceptionally refined investigational techniques would be required to prove their essentiality.

The importance of inorganic chemical elements to the growth and development of living organisms, other than those commonly considered nutritive, was not fully recognized by scientists until about a half century ago. This applies particularly to the group of trace elements, whose functions in the metabolic processes of plants, animals, and humans have meanwhile been demonstrated to be of fundamental significance. Despite our knowledge of their role in the life cycle of a plant and their involvement in complex enzyme activities, the complete relationship of their functions to the visual symptoms of their deficiency or excess is as yet still fractionally understood.

The determination of the essentiality of the presently recognized group of trace elements has been a difficult task, in comparison with that involving the macronutrients, primarily because of the infinitely small amounts with which one must deal. To study the trace elements, the investigators had to employ culture solutions that were absolutely free of the element under examination. The difficulties encountered were great. For example, the applied basal fertilizers contained, in most cases, a sufficient amount of the trace elements as impurities to vitiate the experiment; and the seed might contain a sufficient quantity of the trace element to satisfy the plant requirements for normal growth. A brief reference to how one investigation circumvented this interference will help one appreciate the formidable problem.

Peterson and Purvis[10] undertook the investigation of the role of molybdenum. They grew several plant species in culture solutions in which molybdenum was eliminated by double coprecipitation with copper sulfide as a carrier. In one test normal corn seeds grew to full-sized plants, although they exhibited some visual signs of deficiency. Corn seeds from two ears, taken from plants deprived of molybdenum, contained 0.01 and 0.015 ppm Mo, while the kernels from a third ear were completely free of Mo. The kernels from this third ear were planted in a Mo-free culture solution, and those kernels which germinated produced abnormal seedlings that died within several weeks.

Although it is convenient to group the nutrient elements into "major" or macroelements and "minor", micro- or trace elements, such grouping does not imply degrees of essentiality, for they are all of equal importance. Thus, an absolute lack of any one of these nutrient elements dooms the plant to failure in achieving normal growth. For example, a lack of Mo causes cauliflower and legumes like clovers and peas to fail, even though the amount of Mo needed to promote normal growth is about 2 ounces to the acre. (An acre, plough depth, of soil is estimated to weigh 2 million lb or 1,000 tons.)

[10] Peterson, N. K., and Purvis, E. R., *Soil Sci. Soc. Proc.*, **25**, 111 (1961).

Visual Diagnosis

For many years after Sachs had published his studies on plant growth in water-culture solutions, plant physiologists believed that only the six or seven nutrients he had employed were all that was required by plants for normal growth. However, chemical analysts frequently reported that they found many more chemical elements—as many as 60 in some cases—in the ash of many species and in plants from different localities. In fact before World War II E. Bobko, a Czech scientist, hypothesized that practically all the chemical elements of the Periodic Table are required by plants, although some may be needed in infinitely small amounts. Plant roots absorb and accumulate many elements present in the root zone, many of which seem as yet unimportant for their nutrition but are beneficial in other ways.

The visual method of diagnosis relies on the development of well defined and often characteristic symptoms of the deficiency or excesses. The effects produced are most clearly seen in the leaves. The symptoms are established by growing plants in solution cultures or in pure quartz sand under strictly controlled conditions of nutrition.

Although visual symptoms were at one time the chief diagnostic tool, much has since been learned about trace elements that gives more precise but still incomplete information about the status of these elements. Chemical procedures and new instrumentation permit rapid automatic assessment of the quantity of an element in plant tissues. The specialist in this field is now better able to make fertilizer recommendations, provided that he stays within the admittedly narrow limits of the proven knowledge.

5

DEFICIENCY SYMPTOMS

Modern Depletion Factors

Unquestionably, interest in trace elements is rapidly expanding among soil and crop scientists, the fertilizer industry, and the general public. Greater emphasis seems to be placed on the use and effects of micronutrients than on that of the macronutrients (the so-called major and secondary elements). This is not easily understood, since about 80% of the arable soils of the United States are sufficiently supplied with trace elements, and for the immediate future, at least, the need for supplementing that supply is not urgent. Until quite recently it has been generally assumed that, except for some specific soil areas or speciality crops, the supply of trace elements was adequate for producing satisfactory crop yields, provided that the macronutrient needs were met. The soil's supply of trace nutrients, it was assumed, was being furnished by several sources: crop residues, farm manures, impurities in rainfall and fertilizer compounds, pesticides, the normal weathering of mother rock, and the disintegration of soil particles.

New factors, however, are altering this supply situation and creating localized deficiencies. Among the more obvious of these factors may be listed the following: agricultural practices which intensify the productive capacity of soils; growing two crops per year on the same soil, which seriously depletes the supply of micronutrients, particularly if the soil is a light, sandy type having a low content of them to start with; the accelerated decomposition of the soil's content of organic matter, which is regarded as a storehouse of both macro- and micronutrients; and improved crop varieties having a greatly increased capacity to produce higher per acre harvests if properly fed, but which also hasten the depletion of the soil's reserves of plant nutrients including trace elements. In other words, a supply of plant food that suffices for the production of a 5–10 ton per acre yield of tomatoes will be inadequate for a 20-ton crop yield. Furthermore, since essential nutrient elements are all of equal importance, a deficiency of any particular one can limit the yield.

Another factor of recent origin that contributes to the diminishing native soil supply is the change in the concentration of the current type of fertilizer materials. The components of the NPK basal fertilizers generally demanded are refined to a degree that removes most of their impurities, which include a certain amount of trace elements. Formerly, the general type of mixed NPK fertilizer was formulated with low analysis materials, such as 16–20% P_2O_5 superphosphate, 40–50% K_2O potash, or at times "manure salts" which contained magnesium in addition to potassium. Many of the fertilizer grades also had a substantial amount of organics in them such as cotton seed meal, bone meal, fish meal, and dried blood, which added liberal amounts of trace elements to the fertilizer. The superphosphate furnished, besides phosphorus and numerous trace elements, an amount of sulfur equal to the phosphorus. In present day formulations concentrated 46% P_2O_5 superphosphate and phosphoric acid are employed to furnish the required phosphorus and they both completely lack trace elements, and the superphosphate may have about 1% of sulfur. While farmers used to depend on horses and mules for power and utilized their manure as fertilizer, which carried a variety of trace elements, today's farmer, operating larger farm units, has replaced animal power with machines powered by gas or electricity and replaced low analysis mixed fertilizers with highly concentrated chemical fertilizers.

Crop yields on American mechanized, commercial farms are increasing at an estimated rate of 4–7% per year. The increased harvests drain off the soil's supply of trace elements and also some of the secondary elements (calcium, magnesium, sulfur). Another factor adding to the problem of nutrient supply is the growth characteristic of some crops, which may create a temporary or an acute shortage. For example, the corn crop absorbs more than 75% of its nutrients over a period of 40 days; this uptake causes a strong stress in several localized areas within the soil over a brief time. The increased requirement and intense demand for nutriments often result in the soil's failure to keep up a steady supply and makes it necessary to apply supplemental amounts of trace and secondary elements.

It obviously would help if a nutrient deficiency could be determined before it became too late to prevent a drop in yield. This suggests the use of soil and plant diagnostic techniques as the logical answer. Unfortunately, up to the present time, most routine soil and plant tissue tests are not accurate enough to measure the status of trace elements in a soil. Nutrient deficiency symptoms will often show up in so-called "hunger signs" in the leaves of the plant. Although such symptoms have been identified and catalogued and can be helpful to a plant specialist, when they do appear it is usually too late to do much to prevent a sharp decline in the yield of the affected crop. Correction of the situation in the particular area affected is justified, however, because it could benefit the crop following in the rotation.

PROPERTIES OF TRACE ELEMENTS; BIOCHEMICAL RELATIONSHIPS IN PLANTS AND ANIMALS

6

IRON

Iron is definitely established as an essential micronutrient, required by all plants and animals. Of all the trace elements, iron has the distinction of being the most abundant in soils and plants. It ranks fourth in abundance (about 5%) among the chemical elements in the earth's crust or lithosphere, after oxygen, silicon, and aluminum. Elemental iron is found in the earth only occasionally and then in the form of meteoric rock. However, iron combines readily with other chemical elements and is a common constituent of many hundreds of minerals among which are the oxides, for example, hematite (Fe_2O_3), magnetite (Fe_3O_4), the ferrous carbonate (or siderite ($FeCO_3$)), and ferric carbonate ($Fe_2(CO_3)_3$), and the sulfide, pyrite (FeS_2), limonite ($Fe_2O_3 \cdot 3H_2O$) may be formed through hydration and oxidation of siderite.

Iron is present in many primary minerals such as hornblende, biotite, and chlorite, which are decomposed by weathering and chemical reactions into secondary minerals, such as oxides and hydroxides.

The element, Fe, has an atomic number 26. It shows a maximal oxidation state or valence of $+6$, but generally only the $+2$ and $+3$ states are common. An example of a compound resulting from vigorous oxidizing conditions is barium ferrate, $BaFeO_4$, but the 6 valence compound is rare. Compounds such as Fe_3O_4 are considered as having a fractional oxidation state, or as ferrosoferric oxides which can be designated as $FeO \cdot Fe_2O_3$.

In the $+2$ state iron occurs chiefly as ferrous ion, Fe^{2+}, which easily oxidizes to the ferric, Fe^{3+}, ion upon exposure to the air. Acid solutions of ferrous salts can be maintained for long periods, since the rate of oxidation by O_2 is inversely proportional to the H^+ ion concentration. If a cation is added to ferrous solutions, ferrous hydroxide, $Fe(OH)_2$ is precipitated, which, upon exposure to the air, will turn brown, recognized as being the compound ferric hydroxide, $Fe(OH)_3$. Pure ferric hydroxide has, insofar as is known, never been prepared synthetically.

Iron in the 3 valence or $+3$ oxidation state exists as the colorless Fe^{3+} ion. In both the $+2$ and $+3$ oxidation states iron tends strongly to form complexes; for example, Fe^{3+} combines with the thiocynate ion $(SCN)^-$ to form $Fe(SCN^{2+})$, and with cyanide ion, CN^-, both Fe^{2+} and Fe^{3+} form various complexes.

Fe in Soils

In the soil ferric ion is soluble in the pH range, 3 to about 5, and is available to plants; up to pH 7 or slightly above, ferrous ion takes over; and up to approximately pH 8, ferric and ferrous humates are still soluble. The two humates appear to be readily soluble at nearly all pH levels and are present in small but sufficient amounts for plant uptake wherever humus is present. Absolute iron deficiency, say the authorities, is rare in nature and seems more likely to develop under man's artificial conditions or where sodium is present, causing a pH level of above 8. The deficiency of iron occurring in highly calcareous soils should be attributed to physiological causes in most cases, rather than to the result of an actual lack of iron. Availability of iron for plant utilization increases with acidity but is depressed by phosphates and lime. A heavy concentration of ferric ions in the soil solution can be toxic to most plants.

The element Fe is very abundant in soils, the amount ranging from 200 ppm to at least 10%. The soils termed "ferruginous" contain a high percentage of iron, which gives the soils a reddish color. Leached, acid sands contain the least amount of iron. The iron content of a large number of United States soils ranges from 0.23–11.2%. It should be emphasized that no acceptable chemical test for determining iron availability has as yet been approved officially in the United States[1]. Soil scientists point out that the controlling factor in plant uptake of the element is not the total amount of iron in a soil but only that portion that is available. Many factors operate in the soil to reduce or restrict the availability: a high pH level; bicarbonate ions in the soil solution or in irrigation water; the presence of substantial amounts of magnesium carbonate and calcium carbonate; excessive amounts of PO_4 ions; high levels of copper, manganese, molybdenum, vanadium, and zinc ions; or deficiencies of potassium or calcium ions; or an actual shortage of the element itself[2]. If one or more of these restraining factors is present, adding iron salts to the soil will not greatly help to correct the situation, since the same factors that reduced the availability in the first place will continue to exert the same effect. The introduction and development of chelated iron products have been helpful in overcoming this problem. Also beneficial have been foliar sprays of dilute solutions of ferrous sulfate.

Rarely do soils contain less than 2% (20 tons/acre in top 9 in. layer) iron oxide soluble in hydrochloric acid (equivalent to 1.4% Fe). The Broadbalk Field at Rothamsted (England) without manuring for 50 years had 2.38% Fe_2O_3.[3]

McConnell[4] estimated that a typical British fertile soil would range from about 2.38% Fe for a sandy soil to about 4.4% Fe for humus type soils. F. E. Bear[5] reports that typical iron contents of American soils range from 0.42% in sandy soils to about 5.58% Fe_2O_3 for heavy clay types.

Recent investigations in the United States indicate that iron may often be a limiting factor in the optimal utilization of soil nitrogen and phosphorus. Furthermore, the research findings show that iron exerts a more subtle effect within the plant than, for example, phosphorus. It is possible to measure zinc and manganese availability by the amount absorbed by the plant, but it is not possible to classify iron uptake so definitely. Soil chemists will have to collaborate closely with plant

[1] Berger, K. C., "Introductory Soils," The Macmillan Co., N.Y., 1965.
[2] Wallace, T., and Hewitt, E. J., *J. Pomol. Hort. Sci.*, **22**, 131–161 (1946).
[3] Hall, A. D., "Fertilizers and Manures," John Murray, London, 1920.
[4] McConnell, P., "The Agricultural Notebook," London, 1922.
[5] Bear, F. E., "Theory and Practice in Use of Fertilizers," John Wiley & Sons, N.Y., 1929.

physiologists in order to attain a better comprehension of the distribution and behavior of iron in the tissues of living plants. This approach is also suggested for the investigation of copper, boron, and molybdenum ions, even though at present they appear to be less important and less problematic than iron.

Most of the iron in the soils of humid regions is derived from the weathering and chemical reactions of the primary iron containing minerals, hornblende, biotite, pyrite, and chlorite. The iron in these minerals is mainly in the $+2$ oxidation state, Fe^{2+}, and weathering changes most of the ferrous to the ferric, Fe^{3+}, state. In the process of soil formation the iron of the parent rocks is altered by two major conditions:

(1) At low temperatures and high rainfall, an accumulation of organic cover develops (a condition prevalent in our Northeast Region) under which a large portion of bases or cations (Mg, Ca, Na, K) is dissolved and leached away, leaving an acid residue in the top horizons of the newly formed soil. Much of the iron becomes reduced to Fe^{2+}, is dissolved and carried to a lower, less acid horizon where it is precipitated and forms new iron compounds, such as iron hydrates, oxides, and organic iron complexes. The more hydrated iron compounds dissolve more easily than any of the inorganic iron minerals.

(2) When only a thin organic cover is accumulated, but rain is abundant and temperatures are high (a condition characteristic of many tropical areas), the active cations are leached away, leaving an acid soil and favoring oxidation. The iron is converted mostly into hydrated oxides which do not move, while some of the hydrated iron becomes in time dehydrated to the ferric oxide form, Fe_2O_3. The hydrated oxides coat the finer soil particles and impart to the soil mass various hues of red and yellow.

A deficiency of iron very often develops even in acid soils, frequently limiting the growth of such acid-soil-loving species as azaleas and rhododendrons. This condition may result from an accumulation of the heavy metals, Cu, Mn, Zn, Ni—relative to the amount of available iron present.

Lime containing soils are likely to contain an insufficient amount of available iron. On this type of soil, plants may grow but abnormally, being characterized by a lime induced chlorosis. The visual symptoms are yellowish foliage, lack of vigor, and a general unproductiveness. This condition is commonly encountered in the arid Intermountain and Southwest Regions of the United States.

Iron in Plants

Iron is needed by all plants. It is an essential component of the catalyst involved in the formation of chlorophyll. Until very recently iron has been considered a relatively immobile ion in plants. The prevailing viewpoint was that iron deficiencies could be attributed to immobility[6]. Research at the University of California (Davis) has shown that the iron ion is at least moderately mobile in plants[7,8], and that a good deal of correlation exists between the chlorophyll content of leaves and their iron content[9]. Foliar applied iron is distributed from the leaf to which it is applied to young expanding areas of meristematic activity[10]. The iron is translocated by phloem and xylem to the meristem tissue.

Eddings and Brown[11] studied the absorption of $^{59}Fe^{3+}$ by the leaves of different species. They demonstrated that the iron was translocated from treated leaves and found that this activity varied with different species.

Bukovac and Wittwer[12] found that about 10% of the iron applied to a spot on a bean leaf was translocated, but only within the treated leaf. Doney et al[10] using isotopic $^{59}Fe^{3+}$ found that 25% of the iron applied to a leaf was translocated to the apex leaves of the bean plant.

The amount of iron in the leaves of a normal plant will generally average a few hundred parts per million—the amount hardly varying. One authority estimated an amount of 0.035% in dry matter of ryegrass and 0.049% in that of clover grown in the same field. In general, legumes are richer in iron than grasses (see Tables 6.1, 6.2).

The apparent immobility of iron in plants may be attributed to competition from other iron binding compounds. Some investigators think that physiologically active iron occurs as a metallo–organic complex, which acts as an oxygen carrier, or oxidizing catalyst or enzyme. Iron is known to play an important role in a series of respiratory

[6] Brown, J. C., "Iron Chlorosis," *Ann. Rev. Plant Physiol.*, **7**, 171–90 (1956).
[7] Branton, D., and Jacobson, L., "Iron transport in pea plants," *Plant Physiol.*, **37**, 539–545 (1962).
[8] Brown, J., et al., "Evidence of translocation of iron in plants," *Plant Physiol*, **40**, 35–38 (1965).
[9] Jacobson, L., and Ortel, J. J., *Plant Physiol*, **31**, 199–204 (1945).
[10] Doney, R. C., et al., *Soil Sci.*, **89**, 269–275 (1960).
[11] Eddings, J. L., and Brown, A. L., *Plant Physiol.*, **42**, 15–19 (1967).
[12] Bukovac, M. J., and Wittwer, S. H., *Plant Physiol.*, **32**, 428–35 (1957).

TABLE 6.1. PLANT PRODUCTS HAVING LARGE AMOUNTS OF TOTAL IRON (DRY MATTER BASIS)

PLANT	EDIBLE PORTION	IRON† mg/100 gm	PPM*
Spinach	leaf	41	1,750
Mustard	leaf	40	1,540
Watercress	leaves, stem	31	1,399
Carrots	leaves	—	765
Endive	leaves	25	769
Parsley	leaves	27	1,609
Chard	leaves	27	—
Beets	leaves	—	1,932
Beans, garden	seed	—	769
Turnip	leaves	23	2,483
Yams	tuber	—	340
Rutabagas	leaves	—	1,389
Cabbage, Chinese	leaves	20	—
Radish	root	15	825
Lettuce	—	—	4,800
Asparagus	stem	13	979
Potato	tuber	—	107
Broccoli	flower	13	210
Cauliflower	flower	13	—
Tomato	fruit	10	—
Alfalfa	above ground	—	1,000
Clover	above ground	—	1,300

* Source: N.P.F.I. Fertilizer Rev. (1963).
† Various textbooks compilation.
Note: The total amount of iron in plants used for food is less important, for nutritional purposes, than the percentage of ionizable or nutritionally available iron. The available iron in plant tissue, may range from 20–100% of the total iron shown in Table 5.1 above.

enzymes. It is also indispensable as an active element in the plants' photosynthetic process.

Iron may be precipitated in or near the stalk nodes of the corn plant as a result of alkaline soil conditions. Consequently, such iron is prevented from translocating to the leaves where it is needed. This would be considered a deficiency, by visual symptoms, of the foliage, when actually a sufficient amount of iron may be available but is prevented from reaching its goal.

TABLE 6.2. AVAILABLE IRON IN VARIOUS PLANT PRODUCTS (DRY MATTER BASIS)

		IRON		
		TOTAL	AVAILABLE	
PLANT	TISSUE	(mg/100gm)	PERCENT	AMT. (mg/100gm)
Currant (Black)	Fruit	8.3	100	8.3
Fig	Fruit	5.5	96	5.3
Cocoa	Seed	14.3	93	13.3
Onion	Bulb	6.2	87	5.4
Soybean	Seed	8.5	84	7.1
Bean, string	Seed	6.7	83	5.6
Lentil	Seed	8.2	66	5.4
Potato, white	Tuber	3.2	55	1.7
Rice	Seed	2.3	45	1.0
Peanut	Seed	2.0	45	0.9
Spinach	Leaf	41.0	22	9.0

Source: Rockland, L. B., *U.S. Dept. Agr. Yearbook* (1957).

Iron is an activating element in enzymes. Many enzymes consist of two parts; one part is the coenzyme, which consists of a single metal atom, such as Fe. If this metal portion is removed, it leaves the remaining portion inactive, but if the missing metal is restored, the enzyme will resume its normal activity. Examples of enzymes or oxygen carriers in which iron is the metallic coenzyme in the respiratory tissues of living cells are catalase, peroxidase, cytochrome, and possibly cytochrome oxidase.

Iron Deficiency

Iron deficiency is probably the most commonly occurring of the trace element deficiencies. It occurs in many parts of the world and very frequently on calcareous soils and in fruit crops, but quite often on acid soils also. The chief visual symptom of iron deficiency is a chlorotic mottling of the leaf, particularly on young growth. In severe cases all the green color of the leaf may be lost. In tree crops the condition is often followed by dieback and even death of the tree.

6.1   Iron deficiency in grapefruit leaves, showing chlorotic condition; left leaf normal; right leaf deficient. (*Courtesy of Fla. Agr. Expt. Station*)

Iron deficiency in different species is easily recognized by their visible symptoms, which are the most specific and easily identifiable among all of the plant nutrient deficiencies. The earliest manifestation is usually a pale leaf color. The next stage involves a chlorosis of the leaf areas between the veins, which remain green. In severe cases the entire leaf, veins and interveinal areas, turn yellow. The pattern in grasses consists of alternate stripes of green veins and yellow interveinal areas (see Figures 6.1, 6.2, 6.3, 6.4).

6.2 Iron deficiency in avocado, showing chlorotic leaves. (*Courtesy of I. Stewart, Fla. Agr. Expt. Station*)

Iron deficiency in crops poses more of a problem in the western states of our country. Woody plants seem to show symptoms sooner than succulent plants: pines, pin oak, roses, acid loving plants such as blueberries, azaleas, and rhododendrons. Lawns tend to show deficiencies perhaps because of high levels of phosphate and high pH levels. Putting greens made from mixtures of coarse sand and acid peat moss will readily reveal iron deficiency symptoms. An excessive content of manganese can induce iron deficiency—this is the case in Hawaii in commercial pineapple fields.

As previously mentioned, deficiency symptoms do not always mean that a total shortage of iron is present; many times the cause is a lack of available iron, ferrous Fe^{2+}, due to some adverse or physiological condition preventing a supply of the needed iron to the plant tissues—an induced rather than a naturally occurring condition. However, certain open, sandy soils could in fact lack iron.

Although iron is not a constituent element of the chlorophyll molecule, it is an element of the protein molecule believed to be concerned with its synthesis. A failure in the supply line of iron to the leaf results in its inability to produce chlorophyll, and hence, chlorosis. Plants need a continuous supply of small quantities of iron to function normally.

Diagnostic Methods

Each species has its typical chlorosis pattern. Berger and Pratt[13] give the following diagnostic techniques to confirm an iron deficiency:

(1) Apply a ½ or 1% aqueous solution of ferrous sulfate to recently matured leaves. Unless adverse weather conditions intervene, an increase in green color should occur within two weeks if iron deficiency had caused the chlorosis.

(2) Leaf analysis. In most plants the iron content of leaves generally averages in the range of 10–80 ppm if deficiency is present. It is essential to wash off all dust from plant materials with soap solutions or with dilute acid before analyzing for Fe.

(3) Analysis of soil as a means of estimating the status of iron in that soil is not recommended, since the authorities report that tests for

[13] Berger, K. C., and Pratt, P. F., "Fertilizer Technology and Usage," *Soil Sci. Soc. Am. Proc.* (1963).

6.3 Iron deficiency, showing chlorosis in sweet corn. (*Courtesy of P. Westgate, Sanford, Fla., Agr. Expt. Station*)

availability of Fe in soils have proved to be unsatisfactory. Some work in which the soil was first treated with diethylenetriamine pentaacetic acid showed that the water soluble iron was correlated with the Fe deficiency of plants grown in three different soils.

Corrective Measures

USING SOLUBLE IRON SALTS

If it is known that a real shortage of available iron exists in the soil and is causing deficiency symptoms, the addition of iron salts as a supplement to fertilizer will correct the shortage. If, however, the iron is being rendered unavailable in the plant or soil, as previously described, then correction is likely to fail.

Applying iron salts as foliar spray may not always be successful in correcting the deficiency. In Hawaii the regular spraying of pineapples with a solution of iron sulfate is routine. Some good results with foliar sprays of a 3% ferrous sulfate solution were recently reported by workers at the University of California (Davis).

Krantz and associates corrected iron deficiencies in grain sorghum with foliar spray applications[14]. The chlorosis of the sorghum occurred in Tulare and Fresno counties during the summer of 1960. Soil analyses showed an iron deficiency definitely existed, and correction of the chlorosis in greenhouse grown plants was achieved with soil application or foliar spraying.

Applying a 3% ferrous sulfate solution by spraying the foliage gave a remarkable correction of the chlorotic condition within 7 days. In one severely affected field in Fresno County the spray increased yields from 250 lb/acre to 4,000 lb/acre. Similar remarkable results were obtained in Tulare and San Diego Counties with the same ferrous sulfate solution applications.

Soil applications of either ferrous of ferric sulfates were effective correctives in greenhouse studies but the large quantities required (3,200 lb/acre) rules this soil method out as uneconomical. Soil application of the iron chelate Sequestrene 138 at rates up to 128 lb/acre was only slightly effective. Treating grain sorghum chlorosis by soil applications is therefore not recommended.

Preliminary studies by California University personnel favor the following procedures for correcting iron chlorosis:

(1) In mild cases of chlorosis, apply one foliage spray about 25 days after planting. If chlorosis is severe and plants are stunted, two or possibly three sprayings may be required; the first spray is applied 10 days after emergence and the second, about 25 days after emergence. The third spray may be needed if the chlorosis shows signs of coming back.

(2) The solution to be used is approximately 3% ferrous sulfate solution (25 lb per 100 gallons water). A wetting agent, such as X-77, is essential. The spray should be applied at about 35 lb pressure to form a fine mist to give good coverage on the foliage. About 20–50 gallons per acre should be enough.

CORRECTION MADE WITH CHELATING AGENTS

The name "chelate" is derived from the Greek, Chela (a claw), and is applied to chemicals with the ability to surround and protect certain

[14] Krantz, B. A., *Calif. Agr.*, **16** (5), 5, 6 (1962).

chemical elements. The combination of the appropriate chelate with a metal maintains the metal in a soluble and available form. Chelated metals lose their cationic characteristics but retain their activity in the soil or plant.

Nature has many types of chelates in soil. They include humic, citric, tartaric, ascorbic, and amino acids, but these are not always abundant or stable or sufficiently effective. A good example of a natural chelate is the chlorophyll molecule chelating the metal magnesium.

6.4 Iron deficiency or lime-induced chlorosis in apple. (*Courtesy of National Plant Food Institute*)

The first application of a chelating agent in the control of iron deficiency was made by Stewart and Leonard in 1951 to correct iron deficiency of citrus trees grown in acid sands in Florida[15]. They used a synthetic chelating agent, Fe-EDTA. Since then tremendous advances have been made in the use of chelating agents, not only with iron, but also with other trace elements that are deficient.

The Florida cases of iron deficiency represented an actual low content of iron in the soil. Thus, when the same iron chelate was used in California, the first results were disappointing, for the California soils were alkaline, often having an excess of lime and more clay than sand. Under such conditions EDTA is not sufficiently stable and is easily fixed by clay. Research developed other types of chelating agents suitable for alkaline soil conditions, such as EDDHA.

Chelating Agents

At least five main synthetic chelating agents are now available for agricultural and horticultural use, based on the following acids:

EDTA: Ethylenediaminetetraacetic acid.

HEEDTA: Hydroxyethylethylenediaminetriacetic acid.

DTPA: Diethylenetriaminepentaacetic acid.

EDDHA: Ethylenediamine di-o-hydroxyphenylacetic acid.

APCA: Amino polycarboxylic acid.

EDTA and HEEDTA have proved most effective in making iron and zinc available to plants in acid to slightly alkaline soils. DTPA and EDDHA are suitable for use in calcareous soils.

The relative effectiveness of these chelating agents in making Fe available at pH 7 is reported as follows:

$$APCA > DTPA > HEEDTA > EDTA$$

Fe-APCA has been the most satisfactory chelate used on alkaline, calcareous soils, followed by Fe-DTPA[16].

---

[15] Stewart, I., and Leonard, C. D., "Iron Chlorosis: Its Possible Cause and Control," *Citrus Magazine*, **44** (10), 22–25 (1952).
[16] *U.S. Dep. Agr. Yearbook* (1957).

How Chelating Agents Work

The metal chelated ion is bound by the acid molecule through two or more positions within the molecular structure. For example, in the iron chelate the Fe ion is bound so that it cannot free itself to form another compound when treated with a phosphate or hydroxide. Some of the synthetic chelates that bind the iron are quite soluble, but the iron is held in a soluble complex form available to plants as a nutrient ion. The iron and the chelating molecule may enter the plant together and be transported to the leaves, yet the iron will be held and protected from fixation as it travels to the leaves where it is needed. On the other hand, the iron may be taken away from the chelating agent at the root surface by a natural chelator, which has a more powerful attracting and binding capacity than the synthetic chelator, and then is translocated within the plant.

Unquestionably, many organic compounds formed from soil organic matter are natural chelators that bind or complex metal ions, such as Fe, Cu, and Mn, but many of these organic compounds can be readily destroyed by soil microbial forces.

How Chelates are Used

AS FOLIAR SPRAYS

To be effective a foliar spray solution must be able to penetrate the leaf and to be easily translocated to where it is needed in the plant. Also, the spray must not be toxic nor injurious to the plant nor leave toxic residues in the edible portions of the plant. Chelating agents are decomposed by sunlight, and this is an important consideration. Experience with spray applications of chelates has not been favorable to their use.

SOIL APPLICATIONS

A chelating agent applied to the soil must be able to hold the metal ion firmly against efforts from other metals in the soil to snatch it away. It also must resist hydrolysis, microbial breakdown, fixation by clay particles. It must be ready to yield the metal to the plant as required, be nontoxic to humans and plants and be economical to use.

Soil applications have produced more satisfactory results than spray applications.

The rate of application to correct iron deficiency differs with locality and other local conditions. Typical of the suggested rates of soil application for iron chelates are the following:

Fruit trees: $\frac{1}{2}$–1 lb Fe-chelate per tree, depending on species and age.
Ornamentals: $\frac{1}{4}$–$\frac{1}{2}$ lb per woody shrubs.
　　　　　　 2–8 oz per herbaceous plants.
Field crops: 5–10 lb/acre.

Iron Salts Used

Ferrous oxalate may be applied as dust or spray. It contains 30% metallic iron of which half is available if in chelated form. Good results have been reported when used on grasses, ornamentals, vegetables, and some fruit trees. It is applied at 2–4 lb/acre.

Ferrous sulfate, $FeSO_4 \cdot 7H_2O$ contains 20% as metallic Fe, and is generally used for spray, foliar applications.

Ferric sulfate, $Fe_2(SO_4)_3$, is a more acid forming salt. It should be used only for soil applications, where it acidifies soil and supplies iron.

Zinc-iron-ammonium sulfate and ferrous ammonium sulfate are also offered by some purveyors but are not commonly employed.

Iron in Grassland Herbage

In the chemical analysis of herbage for determination of iron content, it is essential to guard against contamination by soil. This is a more significant factor with iron determinations than with that of any other nutrient element[17]. Analyses (other than of alfalfa) showing values greater than 500 ppm suggest contamination. Of course, smaller values may also include traces of soil.

Plants differ a great deal in their content of iron and in their ability to extract iron from soils. Authorities suggest that pasture grasses average 100–200 ppm Fe, pasture legumes, 200–300 ppm, while for alfalfa values as high as 1,000 ppm have been recorded.

[17] Mitchell, R. L., "Contamination problems," *J. Sci. Food Agr.*, **11**, 553–560 (1960).

Iron in Animal Diet

The need for iron by the animal is directly related to the rate of growth or the loss of blood. The animal body is practically unable to excrete iron into the urine and feces. For this reason control over absorption is essential and is accomplished by the intestinal wall. The amount of iron absorbed is related to its need and the amount of available Fe in the diet.

Animal feedstuffs normally contain enough iron to satisfy the requirements in that element, although no unanimity exists regarding minimal levels. Grazing animals undoubtedly take in a substantial amount of iron with their ingestion of soil[18]. The total iron content of feedstuff is not a good measure of the iron contributed by the feed to the animal; the available is the measure.

Dairy cattle grazing pastures contaminated with ferric hydroxide, $Fe(OH)_3$, derived from iron rich irrigation waters have suffered from its ingestion, the minimal injurious dose being about 30 gm Fe/day for an adult animal.

A large part of the iron content of alfalfa and spinach is poorly utilized, while that of soya beans is much more available to the animal metabolism.

Hogan[19] states that the body of an adult steer contains on average 0.013% Fe; that of a 225 lb hog, 0.013% Fe. About 70% of the body iron is in the blood. The remainder is stored in the liver, bone marrow, and spleen. The Fe content of hemoglobin is practically the same for all animals, namely, about 0.335%.

Piglet Anemia

When young pigs are raised in pens without access to the ground, they tend to suffer from iron deficiency. Most animals are born with enough iron reserves to carry them through the nursing period. Anemia in piglets is often associated with an insufficiency of iron in the sow's milk. It rarely occurs in litters farrowed outdoors and raised on pasture grass.

[18] Hodgson, J. F., "Micronutrients in soils and plants," *J. Agr. Food Chem.*, **10**, 171–174 (1962).
[19] Hogan, A. G., *Missouri AES Res. Bull.*, No. 107 (1927).

An effective control of the anemia as recommended by veterinarians is to dose the piglet with a solution of iron and copper salts. Copper is included because it is usually insufficient in the sow's milk, and it helps to mobilize Fe for the building up of the hemoglobin in the animal's blood. The daily dosage should supply 20–30 mg Fe and 2–3 mg Cu per piglet.

Recent findings at Purdue University[20] indicate that a mixture of palatable feed ingredients and ferrous fumarate fed free-choice seemed to be effective as an oral means of supplying iron to anemic piglets, and it compared favorably with an injection of iron (100 mg Fe) given at 3 days of age.

The results for the average 21 day weights and hemoglobin values were as follows: for the iron injection, 12.6 lb and 10 gm%; for the oral iron mixture, 12.2 lb and 11.1 gm%, respectively.

The iron mixture contained approximately 6.6% Fe; the fumarate, 33% Fe.

The injected iron was iron dextran.

Rickets

The ingestion of large amounts of iron compounds, by design or otherwise, in the rations of young animals may cause rickets to develop. Excessive iron may induce the formation of insoluble iron phosphates, which obviously would interfere with the normal absorption of phosphorus and lead to rickets in the animal.

For Further Reading

DeKock, P. C., "Heavy metal toxicity and iron chlorosis," *Ann. Botany*, **20**, 133–141 (1956).

DeKock, P. C., "Iron nutrition of plants at high pH," *Soil Sci.*, **79**, 167–175 (1955).

Holmes, R. S., and Brown, J. C., "Chelates as correctives for chlorosis," *Soil Sci.*, **80**, 167–179 (1955).

Leonard, C. D., "Chelates in Florida citrus production," *Farm Chem.*, **122**, 48–51 (1959).

Olney, V. W., "Symposium on the Use of Metal Chelates in Plant Nutrition," ed., Wallace, A., pp. 42–44, 54–55, 1956. The National Press, Palo Alto, Calif.

[20] *Purdue Univ. Res. Progress Rept.*, No. 150 (September, 1964).

"A Decade of Synthetic Chelating Agents in Inorganic Plant Nutrition," ed. Wallace, A., 1955, 1962. The National Press, Palo Alto, Calif.

Smith, R. L., "The Sequestration of Metals," The Macmillan Co., N.Y., 1959.

Chaberek, S., and Martell, A. E., "Organic Sequestering Agents," John Wiley and Sons, N.Y., 89, 1959.

McElroy, W. D., and Nason, A., "Mechanism of action of micronutrient elements in enzyme systems," *Ann. Rev. Plant. Physiol.*, **5**, 1–30 (1954).

For details of application–methods, rates, etc., the literature prepared and released by individual manufacturers should be consulted.

7

MANGANESE

Manganese (Mn) is a nutrient essential to the normal growth of plants, animals, bacteria, and fungi. It is present in all plant tissues but is particularly concentrated in the green leaves, shoots, and seeds. In animals it is found particularly in the liver, pancreas, and hair, but is present in other tissues. It is not a very common chemical element, being present at about 0.08% in the earth's crust. Its essentiality as a nutrient was recognized shortly after 1920, following publication of extensive research on its role in plant growth by J. S. McHargue, scientist at the Kentucky Agricultural Experiment Station (AES).

Manganese acts as a catalyst in several important enzymatic and physiological reactions in plants. In excess amounts it can be toxic to plants and may also reduce the availability of iron to plants. Iron is normally absorbed by the plant as the ferric trivalent ion (Fe^{3+}) and is reduced by it to the ferrous divalent form (Fe^{2+}) and utilized as such in the plant's metabolism. In the presence of an excessive amount of Mn the ferric iron remains unusable by the plant, with the consequence that the plant suffers from iron deficiency.

Manganese is generally involved in the plant's respiratory process such as the oxidation of carbohydrate to carbon dioxide and water. This is catalyzed by an enzyme activated by manganese. Besides activating enzymes catalyzing various stages in plant respiration, manganese also activates enzymes which are concerned with the metabolism of nitrogen.

Another important function is the activation of enzymes directly involved in the synthesis of chlorophyll. Mn is quite immobile in plant tissues; this is why the earliest deficiency symptoms are visible in the younger leaves or shoots. Physiologists now accept the evidence which indicates that manganese, together with iron, controls the reduction-oxidation potentials in plant cells during the phases of light and darkness.

Chemical Relationships

Manganese occurs in a number of minerals, the most important being pyrolusite (MnO_2), braunite (Mn_2O_3), hausmannite (Mn_3O_4), and manganite ($MnO(OH)$), and as a trace constituent in some of the primary minerals, hornblende, olivine, biotite, garnet, and augite. In the periodic table it has an atomic number 25 and its atomic weight is 54.938.

In chemical compounds it shows valences of 2, 3, 4, 6, and 7. Its manganous ion (Mn^{2+}) is characterized by a pink color, one of the very few pink ions in chemistry. Mn^{2+} is not a good reducing agent; hence neutral or acid solutions of Mn^{2+} salts can be kept unchanged for a long time exposed to oxygen or other oxidizing agents. If a base is added to a manganous (Mn^{2+}) salt solution, $Mn(OH)_2$ is formed as a white precipitate which, exposed to the air, is readily oxidized to Mn^{3+}. The oxidation potential of Mn^{2+} can be represented as $Mn^{2+} \rightleftarrows Mn^{3+} + e^-$, and is -1.51 volts.

Manganese can exist as manganic ion (Mn^{3+}) only in solids and complex ions. Mn^{3+} is a strong oxidizing agent; it can oxidize itself to tetravalent Mn^{44} and even H_2O to liberate O_2.

In the tetravalent state the principal compound is manganese dioxide (MnO_2), which, when heated with basic substances in air, is oxidized to the manganate ion $(MnO_4)^-$ which is its highest oxidation state.

Hexavalent manganese (Mn^{6+}) is stable in alkaline solution but unstable in acid solution.

In its low oxidation state manganese exists as a cation which forms basic oxides and hydroxides; in its higher oxidation states, for example as $(MnO_4)^-$, it exists as an anion derived from acidic oxides.

Determining the Amount of Mn

A series of manganese oxides as well as many other manganese compounds form in soils. These range from manganous to highly oxidized manganic forms. This explains why it is very difficult to determine precisely the amount that is available to plants. Also present is some manganese associated with soil organic matter composed mainly of plant materials, which must be released by soil microbial agencies before it becomes available to plants. However, some microorganisms convert the available Mn^{2+} to higher oxides which are insoluble and unavailable.

Research workers use a number of methods which give reasonable accuracy in the determination of manganese when present in minute quantities. The spectograph, polarograph, and absorptiometer are all being employed quite successfully for this purpose. The spectrograph is considered the most sensitive method of measuring small quantities of many elements, particularly when an arc or spark is used. When the flame method is used, it is less sensitive than the arc.

A method of estimating trace elements in soils and plant materials known as *bioassay* depends on the effect of these elements on the growth of *Aspergillus niger*. Essentially the method consists of growing this fungus on culture media, which contain all the essential nutrients in adequate amount except the one to be determined. A definite quantity of the soil or plant material such as ash is added to the medium, and the amount of growth the fungus makes after a standard time at a definite temperature is determined from the dry weight of the fungus. This is then referred to curves plotting the amount of growth made by the fungus against known quantities of the element present in the culture medium.

Several methods for determining Mn by means of the absorptiometer are available. Most of them depend on oxidizing the manganese salt to produce permanganate, whose intensity of color is then determined by the instrument. In accordance with the oxidizing reagent used, the methods are known as the periodate, persulfate, and bismuthate. The bismuthate method is particularly recommended for the determination of Mn in soils.

A recently proposed method that seems to correlate well with plot tests uses dilute orthophosphoric acid.

Manganese in Soils

Manganese occurs in soils in various degrees of solubility. The total amount of Mn in soils varies widely from a trace to more than 7%. Some soils in the Hawaiian Islands contain as much as 15% MnO, equivalent to 11.6% Mn. Most arable soils in the United States contain between 0.01 and 0.5% Mn. Some soils in Scotland contain at least 0.02% Mn. The Mn content of soils in different parts of India ranges from 96–1,340 ppm (0.01–0.13%).

Leeper[1] described the chemical reactions of Mn in soils as interesting from three viewpoints, namely (1) it is not sufficiently available in some types of neutral and alkaline soils to maintain the normal growth of plants; (2) on some acid soils plants absorb it in toxic amounts; and (3) the distribution of the various forms of Mn in soils is closely linked with the process of soil formation.

Scarseth and Salter[2] give this interesting picture of the function of the element: "manganese seems to act as a two-handed (double valence) reception committee, of which zinc and copper are also members, to greet the other nutrient ions as they enter the plant cell and to direct them to their respective positions for carrying out their functions in the plant. Another way to describe this is to say that these elements act as catalysts."

Russell[3] summarizes the different aspects of Mn somewhat as follows: cultivated soils usually contain less Mn than uncultivated soils; soil pH and the oxidation-reduction equilibrium of a soil govern the degree of its solubility; Mn is one of the most easily exchangeable cations, particularly under acid conditions; liming reduces the solubility and hence availability, probably as a result of oxidation.

Wallace[4] has these observations: Mn resembles Fe in the way it occurs in the soil in that its oxides are important forms. Both Fe and

[1] Leeper, G. W., *Soil Sci.*, **63**, 79–94 (1947).
[2] Scarseth, G. D., and Salter, R. M., "Hunger Signs in Crops," NPFI, Washington D.C., 1941.
[3] Russell, F. C., "Imp. Bu. Animal Nutr.," *Tech. Commun.*, No. 15 (1944).
[4] Wallace, T., "Diagnosis of Mineral Deficiency in Plants," H.M.S.O., 1943.

Mn are intimately involved in the oxidation-reduction reactions occurring in soils and in compounds of both elements in the form of accretions formed in poorly drained soils. The more highly oxidized forms, for example MnO_2, are but slightly available to plants. Solubility increases with increasing acidity, and in many soils it becomes unavailable to plants at pH levels above 6.2–6.5. Organic matter in the soil and drainage conditions directly affect solubility. Deficiency of Mn is common on calcareous peats, on other soils having a high content of organic matter, and on soils with a high water table.

Manganese occurs in many soils in the form of MnO_2, often in a hydrated form, sometimes as Mn^{3+}, and also as an exchangeable cation in the divalent form (Mn^{2+}). Plants can absorb and use Mn^{2+} but not Mn^{4+}. In agricultural practices it is desirable therefore to maintain a reasonable amount of divalent Mn^{2+}. This is fostered by controlling the form of organic matter in the soil and the conditions favoring the microbial population and its activity.

Acid mineral soils are usually well supplied with divalent Mn^{2+}, sometimes excessively so, whereas calcareous soils are usually insufficiently supplied. Many organic soils are very poor suppliers of divalent manganese to crops, despite a high total content of the element. When Mn deficiency occurs in crops on either of these two types of soil (calcareous and organic), it is more effective to spray the crops with corrective Mn than to apply it to the soil. Some success has been attained by adding sulfur or other acidifying agent to an organic soil in order to decrease its pH below 7; this method is almost as effective as adding a soluble Mn to the soil. Mn deficiency resulting from excessive liming can often be controlled by spraying 1–2 lb $MnSO_4$/acre, but better results are achieved by using 5–20 lb of $MnSO_4$/acre for sprays and applying up to 40–50 lb/acre of the salt to peat soils.

In very acid soils the manganese may become so soluble, it becomes toxic to plants. Because the amount of soil Mn may be either excessive or insufficient for normal crop growth, depending on conditions, it behooves the farmer or adviser to understand the factors controlling the availability of Mn in the soil. The most important soil factor concerning Mn availability is the degree of soil acidity, pH 4–5.5. Decreasing the degree of soil acidity reduces the danger of Mn toxicity which usually occurs at pH 4–5. Liming a soil to raise the pH to 6 plus, will strikingly decrease the availability of Mn, and the danger of toxicity is eliminated. But, if the pH rises above 6, there is danger that a deficiency may occur due to the development of insoluble Mn oxides.

Organic matter in soils has the ability to form various complexes with Mn, thus holding Mn in relatively unavailable form. The more tightly held the complex, the higher the pH is raised above 6. Organic matter may also cause the chemical reduction of Mn oxides.

Waterlogging of soils is another factor influencing Mn availability. It increases the availability of Mn, but if too prolonged, it can lead to toxicity by excessively diminishing the soil's supply of oxygen.

Another factor is that as light intensity decreases, the plant's uptake of Mn declines. As the soil temperature decreases (as under cold, moist springtime conditions) the availability of Mn decreases, at times to the extent of causing a deficiency in the plant.

Reactions in Soils

Of the total manganese present in a soil, only a very small proportion is likely to dissolve in the soil solution, and it is this amount that is available to the plant. As this Mn is absorbed it is replaced by exchangeable Mn held by the colloidal complex of the soil. A number of solid manganese compounds may occur in the soil, the element being divalent, trivalent, and tetravalent, but the divalent (Mn^{2+}) is soluble and is the form which replenishes the absorbed portion. Although some workers believe the practically insoluble Mn^{3+} and Mn^{4+} forms contribute to the available manganese, the consensus is that soluble Mn^{2+} constitutes the major source of supply to the plant. However, it is known that the equilibrium of exchangeable and absorbed manganese is not stable, and oxidation and reduction of the soil manganese means changes in valency, that is Mn^{3+} and Mn^{4+}, can be formed from Mn^{2+}, so that the amount of available Mn can vary accordingly. Alkalinity and aeration also favor oxidation and influence these changes in valency, whereas acidity and lowered oxygen supply increase the amount of divalent Mn^{2+}. Soil organisms are also capable of affecting the oxidation of soil Mn, being most effective when the soil pH is from 6.0–7.9.

Iron salts may also influence the amount of soil Mn available to plants. For example, adding ferrous salt to the soil induces the reduction of Mn^{3+} and Mn^{4+} to the divalent state, in which it is more available to plants*. Some of the symptoms of manganese toxicity may be due to manganese induced iron deficiency. Research at Rutgers University indicated that the ratio of Fe to Mn in the nutrient medium of soybean plants and perhaps of other plant species also, should be between 1.5 and 2.5 to assure best plant growth. If the ratio is above 2.5, symptoms of Fe toxicity (or Mn deficiency) would develop; below 1.5 the plant would suffer from Mn toxicity (or Fe deficiency).

Manganese: Plant Relationships

One fact about manganese in plants stands out: the wide variation in manganese content in plants grown under normal conditions. Perhaps this is due to the similarly wide variation in the availability of soil manganese. The manganese content of plants varies much more than for any other micronutrient. Beeson[5] reports a range of from 14–936 ppm in alfalfa, and from 79–510 ppm for red top grass. The National Science Foundation found in a study of 773 analyses of corn an average content of 6 ppm, and for oats and wheat, an average of 19.5 and 24.9 ppm, respectively. Plants of the same crop can differ in their manganese content. All chlorophyll tissues have the highest concentration of the element. Hale and Heintze[6] state that the total manganese in green leaves normally varies from 30–500 ppm of the ovendry material. Olsen[7] determined the Mn content in the leaf blades of a number of plants growing on Danish soils at a wide range of pH values and found a close correlation between soil pH and Mn content; above pH 7 he found values less than 100 ppm, whereas on the most acid soils the leaves contained more than 1,600 ppm on the dry basis. Plants on swampy soils had a high content of manganese even at pH values above 7. The evidence suggests that the Mn content of naturally growing plants will vary with the individual species, the pH, and the water content of the soil. The availability of manganese depends largely on soil pH and the oxidation–reduction potential, hence, it is to be expected that plants growing on acid and on waterlogged soils

---

* For a detailed description of factors influencing absorption of Mn and effects on plants, see Stiles, W., "Trace Elements in Plants," 3rd edition, Cambridge Univ. Press, 1960.

[5] Beeson, K. C., *U.S. Dept. Agr. Misc. Pub.*, **369** (1941).

[6] Hale, J. B., and Heintze, S. G., *Nature*, **157**, 554 (1946).

[7] Olsen, C., *Biochem. Z.*, **269**, 329–348.

TABLE 7.1.   MANGANESE REMOVAL BY SPECIFIED CROPS

CROP	YIELD PER ACRE	lb Mn REMOVED PER ACRE
Corn		
Grain	150 bu	0.09
Stover	4.5 tons	1.50
Wheat		
Grain	40 bu	0.09
Straw	1.5 tons	0.16
Rice		
Rough	80 bu	0.08
Straw	2.5 tons	1.58
Alfalfa	4 tons	0.44
Cotton		
Seed and Lint	1,500 lb	0.11
Peanuts		
Nuts	2,500 lb	0.01
Sugar Beets		
Roots	15 tons	0.75
Tobacco		
Leaves	2,000 lb	0.55

Source: *N.P.F.I. Fertilizer Rev.* (1964).

will have a relatively high Mn content. Table 7.1 gives some data on manganese uptake from specified soils.

Mn Deficiency in Soils

Although manganese availability is reduced in the more alkaline or calcareous soils, many areas in the more humid parts of the country often report deficiencies on organic soils. At present manganese deficiencies have been recorded in 25 states. The crops most affected by deficiencies appear to be oats and soybeans. The minimal level of Mn in healthy oats at the flowering stage is 14 ppm and below this quantity the crop is affected with the so-called "gray speck disease." The condition also affects other cereal and grass crops. The affected plants develop first a grayish lesion on the base of the leaves, which then enlarges and becomes bright yellow or orange along the edge of the leaf during the "halo stage." The tissue within the lesion dies and becomes gray, the leaf top remains green, the base dies and is followed rapidly by the chlorosis of all green tissue.

Symptoms of deficiency in other plants have received popular descriptive names; for example, in cereals gray speck or white streak or dry spot; in field peas, marsh spot; in sugar cane, streak disease or Pahala blight; and in spinach and field beans, yellow disease.

French scientists, using the culture solution technique, were able to induce Mn deficiency by having an excessive amount of iron in the nutrient solution. Optimal growth of test plants was obtained with an Fe : Mn ratio of 5 : 1 to 100 : 1. A prerequisite for this relationship is that the Mn content should be higher than 0.1 ppm and does not exceed 25 ppm. Solutions of $MnSO_4$ increased the Mn content of soybean, corn, wheat, and oats plants.

The fixation of added Mn, converting it into an unavailable, nonexchangeable form in many neutral or alkaline soils, occurs rapidly after liming, and therefore, the pH of such soil should be reduced at the same time to prevent a manganese deficiency. Liming of a strongly acid soil to an alkaline condition is a common cause of manganese deficiency. An acid soil having a low content of available Mn should be limed to a pH 5.7–6 value. It has been shown that liming a slightly acid organic soil that does not need the lime will result in depressing the normal growth of onions, potatoes, beets, cereals, legumes, and some tree crops.

An imperfectly drained mineral soil with a high content of organic matter may naturally overlime itself in some places. Where conditions fluctuate between poor and good drainage, a manganese deficiency may develop between seasons, particularly if the soil contains a substantial amount of calcium and magnesium ions. French sources believe that Mn is fixed by adsorption on the surface of calcium carbonate particles.

Previous mention has been made regarding the redox potential of a soil; the lower this potential, the more marked is the reducing power of the soil, with the result that the manganic oxides (Mn^{3+} and Mn^{4+}) are converted into the soluble oxide (Mn^{2+}) which can be absorbed by plants. The reducing properties of a soil increase with saturation with water, with reducing soil microorganisms and their decomposition products, and with the reducing decomposition products of the organic matter. How the organic matter affects the conversion of Mn^{2+} to the tri- and tetravalent forms is not yet completely understood. But many researchers have shown that the presence of a certain amount of organic matter is needed to bring about manganese deficiency of the plants. The mechanism of the microbial oxidation of manganous compounds to insoluble manganic oxides is also not quite clear, although Australian workers have proved that some soil bacteria are capable of oxidizing Mn^{2+} to Mn^{3+}. The fundamental principles of the manganese cycle in the soil are thought to proceed as follows:

Manganous Mn $\rightarrow$ colloidal hydrated $MnO \cdot MnO_2 \cdot (H_2O)_3 \rightarrow MnO_2$

This reaction is reversible if a reducing agent is added:

Colloidal $(MnO)_x (MnO_2)_y \cdot (H_2O)_3 \rightarrow MnO + _y(MnO_2) + z(H_2O)$

Dehydration will cause the hydrates to be reduced by increasing the concentration of hydroxyl ions by means of lime and then mulching to prevent further dehydration.

Corrective Measures

A common method of correcting Mn deficiency in soils is by applying a manganese salt, usually manganese sulfate, $MnSO_4$. The amount to apply is determined by soil type and its known fixation capacity. One general recommendation is to add 50–100 lb/acre of $MnSO_4$ to mineral soils, where the deficiency has been caused by excessive liming. Generally, it is customary to apply:

On soils slightly acid to neutral, 50–100 lb $MnSO_4$/acre.
On soils neutral to slightly alkaline, 100–200 lb $MnSO_4$/acre.
On soils strongly alkaline, 200–400 lb $MnSO_4$/acre.

Another method is to apply $MnSO_4$ by spraying it on the foliage. A very recent study by R. G. Hoeft at the University of Wisconsin[8] indicates that foliar application of manganese on oats and soybeans is the best remedy for correcting Mn deficiency which occurs on Wisconsin's alkaline soils. Hoeft suggests spraying $MnSO_4$ or manganese chelate (MnEDTA) solution as soon as the plant leaves can intercept enough of the chemical.

[8] *Agrichem West* (April, 1968).

TABLE 7.2 SOIL SCIENCE SOCIETY SURVEY ON MANGANESE DEFICIENCIES IN U.S. AND CORRECTIVE MEASURES*

STATE	CROP	CHARACTERISTICS OF DEFICIENT SOILS	DEFICIENT LEVELS, ppm SOIL	DEFICIENT LEVELS PLANT, ppm (OVENDRY)	POUNDS OF Mn/ACRE CORRECTIVE	CARRIER
Illinois	Soybeans, oats.	Sandy soil, pH above 7, most common when June is wet and cool.	—	—	2–3	MnSO$_4$
Indiana	Soybeans, oats, wheat, corn.	Dark sandy soils. Heavy depressional soils.	—	—		MnSO$_4$
Iowa	Oats.	Calcareous, high in organic matter; poorly drained soils.	—	—	Not recommended treatment with MnSO$_4$. Foliar spray is effective.	MnSO$_4$
Michigan	Market garden crops, soybeans, beet sugar, oats, beans, potatoes, sudan grass.	All yellowish sands, humic gray soils, and poorly drained soils. pH 6.5.	—	25	5–20 lb/acre on organic soils.	MnSO$_4$; Mn$_2$O$_3$ if mixed with acid forming fertilizer.

TABLE 7.2 (continued)

Ohio	Soybeans.	Lake bed soils with high content of organic matter.	40	20–40	5–10	$MnSO_4$
Florida	All vegetables; citrus.	Alkaline sands or everglades peat. Some limed soils having pH 6.5–7.	1	8–10	3 2.4 0.8	$MnSO_4$ Mn_2O_3 Manes
Georgia	Sweet potatoes.	Sandy soils of coastal plains.	<17 when pH is above 6–6.2	—	9	$MnSO_4$
Maryland	Corn, soybeans, small grain.	Light sandy soils and black soils high in organic matter. Sandy soils with pH values above 6–6.2 and high P_2O_5 values.	<17 when pH is above 6–6.2	—	8	$MnSO_4$

TABLE 7.2 (continued)

STATE	CROP	CHARACTERISTICS OF DEFICIENT SOILS	DEFICIENT LEVELS, ppm SOIL	DEFICIENT LEVELS PLANT, ppm (OVENDRY)	POUNDS OF Mn/ACRE CORRECTIVE	CARRIER
North Carolina	Soybeans, peanuts.	High organic matter, poorly drained, pH above 6.2.	19	—	8	$MnSO_4$
South Carolina	Cotton, soybeans, snapbeans.	Coastal plain and Piedmont soils derived from acid rocks. Sandy soils low in organic matter with pH 6.3 or above.	—	—	16	$MnSO_4$
Virginia	Soybeans, peanuts.	Overlimed sandy soils of coastal plain.	—	—	8	$MnSO_4$
W. Virginia	Cauliflower.	Organic soils.	—	—	16	$MnSO_4$

TABLE 8 (continued)

Kentucky	Soybeans.	Old alluvium from tributaries and backwater from Ohio River, generally pH above 6.5.	—	20 in leaf tissues	0.73–1.1	—
California	Fruit trees, nut trees.	Highly calcareous.	—	15–20	500–1,000 as foliar spray. 1–5 lb $MnSO_4$/acre.	$MnSO_4$
Oregon	Fruit trees.	Peat soils, Willamette Valley, The Dalles Milton, Freewater.	—	—	3–5 lb Mn/100 gal. per year.	$MnSO_4$
Utah	Fruit trees.	Calcareous and non-calcareous soils.	—	—	2–4 lb Mn/tree per year.	—

* This survey was released by the Soil Testing Committee of the S.S.S. of America, in 1965, and included boron, copper, iron, zinc, molybdenum, besides manganese deficiencies.

71

Other manganese compounds used to correct deficiencies are:

Inorganic: manganese oxide, oxysulfate, fritted Mn, $MnNH_4PO_4 \cdot H_2O$.
Organic: Rayplex Mn; Sequestrene, Na_2Mn.

On alkaline soils an acid forming material, usually a fertilizer such as $(NH_4)_2SO_4$ is applied to prevent the fixation of manganese. Often weedkillers and insecticides are mixed with the $MnSO_4$ and this practice is generally successful with the highly soluble, pure grades. The poorer the solubility and purity, the greater the likelihood that mixing will not be possible and that a precipitate will be formed[9].

Manganous oxide is not water-soluble but is soluble in weak citric acid and therefore is slowly available to the plant. When applied to the soil, it does not become converted into the unavailable manganic salt. Some research in Europe is concerned with applying manganous oxide to the soil to correct manganese deficiency for 3–4 years so as to obviate the need of annual spray applications.

Mn Deficiency in Plants; Symptoms

Some of the visual symptoms, briefly described as occurring in the plants listed below, have been published by numerous authors (see Figures 7.1, 7.2, 7.3).

SOYBEANS

The occurrence of Mn deficiencies in soybeans and oats has been increasing in the Midwest. Deficiency symptoms are usually expected in soybeans when the total Mn in the plants drops below 20–25 ppm. Symptoms are: pale green irregular areas develop between the main veins of leaves, and in time most of the affected areas turn brown (see Figure 7.2).

ALFALFA

Plants become stunted; midstem leaves show first marginal paling, then pale brown or buff spotting near the margins; the younger leaves become distorted at their tips; in time leaf tips become palish yellow-green or gray-green; the leaf margins become brown and irregularly speckled.

[9] Toms, A. M., *Fertilizer Feeding Stuffs J.* (March 20, 1968).

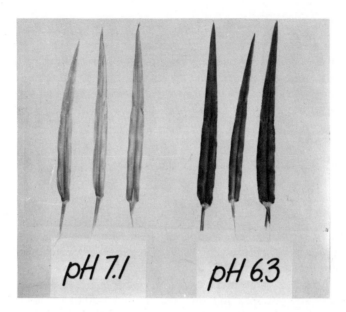

7.1 Manganese deficiency in barley leaf, showing chlorotic areas between veins and the influence of soil pH on the availability of manganese.

POTATO

At first the lower stem at the base of petioles shows dark brown streaks. Chlorosis develops in the interveinal areas on the lower leaves, while the veins remain green.

CAULIFLOWER

The margins of the middle and older leaves show a marked forward cupping. A marginal paling develops and gradually spreads between the major veins, followed by dark brown spotting in pale areas and severe marginal and interveinal crinkling and distortion.

CARROT AND LETTUCE

Stunting of plants; foliage takes on palish color and dull yellow around leaf margins: in carrots this is followed by bronze speckling and scorching along leaf margins.

TOBACCO

Young leaves become palish; tissue between veins is light green to almost white while the veins remain greenish. Leaves appear checkered, due to the contrast between the green veins and the interveinal areas. Later the interveinal tissues show dead spots over the entire leaf. Stunting of the plant may also occur.

SMALL GRAINS

Grayish areas show up near the base of the younger leaves and slowly become larger and become yellowish to yellow-orange.

BEANS

Veins remain green, while interveinal areas are yellow and often develop a series of brownish specks parallel to the veins until finally the whole leaf surface yellows.

When Mn deficiency symptoms appear on soybeans, Illinois authorities suggest spraying the crop with 10–20 lb of $MnSO_4$/acre, preferably in concentrations of not more than 1 lb to 5 gallons of water. Soil applications of $MnSO_4$, particularly after the crop has been established, have usually proved ineffective.

Toxicity Symptoms

A general appraisal is as follows: deformed leaves, chlorotic areas, dead spots, stunted growth, and depressed yields. In cotton plants the symptom called 'crinkle leaf" occurs; the name applies because leaves pucker at edges which cup downward and the edges may become ragged, thicker, and rather more brittle and often chlorotic. Mn toxicity in soyas is much like the pattern in cotton.

Manganese in Animal Metabolism

During the past three decades remarkable advances have been achieved in our understanding of nutrition, based principally on the results of biochemical investigations of plants and animals, including man. Progressively, investigators have turned their attention from carbohydrates, fats, proteins, water, and salts, that more easily studied bulk

7.2 Manganese deficient soybeans, showing light colored tissue between leaf veins that tend to remain green. (*Courtesy of Soybean Digest*)

of body mass, to enzymes, hormones, vitamins and metals, which are present in minute quantities but exert powerful influences in plants and animals alike. One reason for the delay in the investigation of problems involving these essential elements was because instrumentation was either not available or not sensitive enough to deal with quantities present in concentrations as low as thousandths and millionths parts of grams. In recent years new techniques have been devised which have enabled researchers to study substances present in trace quantities as they participate in cells and tissues and function in living systems.

7.3 Manganese deficiency in orange leaves. (*Courtesy of Citrus Expt. Station, Florida*)

Biochemists recognize that manganese is essential for the proper formation of bone and for the development and function of the reproductive system, in fact that it is necessary for the normal growth of the whole animal. Although biochemists have well established the needs for this metal in these essential roles, they have not yet unravelled the mechanisms involved in these various vital functions. It is believed the missing link may lie in the activation of certain enzymes by manganese (Mn^{2+}); for example, bone phosphatase needs Mn^{2+} as a cofactor. Muscle adenosine triphosphatase, certain peptidases, and choline esterase may depend upon Mn^{2+}. Its participation in these and other enzyme systems can account for the animal's dependence upon it.[10]

A great deal of attention has been devoted to the function of manganese in the metabolism of poultry, both mature and immature birds; these birds need feeds containing 30–40 ppm of Mn. The form in which manganese is added to prevent perosis or slipped tendon and for normal growth of chicks seems to be immaterial. Fowls seem to need larger amounts of Mn than mammals. Corn is a poor source of Mn and hence is considered inadequate for poultry.

Mn deficiency is as yet unknown in man, although it may be required as in other mammals. Tea, legumes, leafy vegetables, fruits, nuts, and sea food are good sources of Mn, and their inclusion in human diets may provide the needed manganese.

The Mn content of the animal body is small. It is stored chiefly in the liver but is present also in the pancreas, kidneys, sex organs, skin, muscle, and bones. According to Fearon[11] the range of the amount is from 0.01 mg of Mn/100 gm of fresh tissue in muscle to 0.2 mg in liver. Fresh bone contains from 0.03–0.1 mg Mn/100 gm.

Functions of Mn in Animal Body

The published information is not too precise. Mn seems to affect rate of growth, skeletal metabolism, ovulation, and development of the fetus. It is definitely involved in reproduction. Norris[12] states that it is needed for the growth of chicks, the development of chick bones, the maintenance of egg production, strength of egg shell and hatchability of eggs, and is associated with the activities of certain enzyme systems.

Toxic Dose of Mn

The toxic dose of Mn for animals greatly exceeds the nutritional requirements. Underwood[13] reports that as much as 646 ppm of Mn

[10] Mallette, F., "Biochemistry of Plants and Animals," John Wiley & Sons, N.Y., 1960.
[11] Fearon, W. R., "An Introduction to Biochemistry," Heinemann, Ltd., London, 1946.
[12] Norris, L. C., "Functions of Manganese in Poultry Nutrition," Cornell University.
[13] Underwood, E. J., Nutr. Abstr. Rev., 9 (3), 515–534 (1940).

fed by some researchers and 1,000 ppm fed by others in the diet of chicks caused no toxic effects, but a ration of 4,800 ppm was too toxic to young chicks. Hence, there is a wide margin of safety, at least in chicken feed as currently processed.

Manganese in livestock feeds has been a major study in the Animal Science Department at the University of Florida. Laboratory studies have shown that Mn is required in the diet of rats, rabbits, mice, and chicks, and doubtless in man's diet also, but up till now it has not been proved that depriving man of Mn is harmful. Feaster[14] reports the following items:

MN IN BEEF AND DAIRY CATTLE

Cattle have shown poor growth, leg disorders, poor fertility, and frequent abortion when grazing on pastures grown on certain sand and peat soils low in Mn. These disorders were prevented by feeding 2 gm Mn per day per cow.

In an experiment with dairy cattle, cows were fed hay low in Mn plus corn silage and a grain mixture of corn and corn products which contained an overall Mn content of 3–5 mg/lb of diet. The cows were slow to show signs of estrus and had a great number of calves with weak legs. An addition of Mn to the diet prevented these disorders.

MN IN SWINE

Mn deficiency in swine diets results in these symptoms: reduced growth of the bones with shortening of the leg bones, enlarged hocks and muscular weakness; back fat was increased and the estrus cycle became irregular or completely lacking. Fetuses in pregnant sows often failed to survive and the pigs that were born were small and weak; udder development was poor and the milk production was almost nil.

The leg weakness was diagnosed as slipped tendon, like that occurring in Mn deficient chickens.

The amount of Mn required in the diet of swine has been variously estimated; some workers suggest 5–6 mg Mn/lb of diet, others think this amount is insufficient. The age of the pig, the amount already stored in his tissues, the level of Ca and P in the diet are factors which need to be considered when calculating the amount needed.

[14] Feaster, J. F., *Sunshine State Agr. Res. Rept.*, 18 (October, 1963).

Hens in the second year of egg production need 35 mg of Mn/lb of feed, and pullets require not more than 20 mg for egg production.

Knowledge of the mineral nutrition of farm animals is far from complete, and it is a disservice to the animal industry to evince any pretence to the contrary. More research and experience are needed to extend our knowledge.

Mn and Fertilizer

Ammonium sulfate has a beneficial effect on the availability of soil Mn to plants. This is in sharp contrast to the influence of sodium nitrate. The effect is entirely indirect, since it is associated with changes in pH which result from the uptake of NH^+_4 and NO_3^- by the plants. Optimal manganese absorption relative to the pH of the soil was found to occur at pH 6 and 7. Ammonium sulfate creates an acidic condition in the soil, whereas sodium nitrate tends to leave it alkaline. It has already been shown that Mn^{2+} is available and exchangeable in an acid soil.

Superphosphate exerts a beneficial effect on the uptake of applied Mn when both fertilizer materials are mixed, due to the precipitation of manganese as manganous phosphate. This phosphate retards the oxidation of the manganese and slowly releases small quantities of divalent Mn^{2+}. The superphosphate surrounding the manganese causes a slight acidity as it dissolves, which is helpful in preventing a rapid oxidation of the added manganese.

Manganese sulfate ($MnSO_4$) in various grades of purity (23–25% Mn) and manganous oxide (MnO) commercial grade (about 48% Mn) are the popular materials used as fertilizers to supply manganese.

It must be obvious from previous discussions that if manganese sulfate is applied broadcast on an alkaline soil, there is the danger that the Mn^{2+} will be oxidized and rapidly changed into an insoluble form. If it is applied to such a soil, it must be mixed with an acid forming fertilizer (e.g. $(NH_4)_2SO_4$) or with a phosphate.

On acid soils in which Mn is deficient (because they have low amounts of the element), the manganese may be applied broadcast or in bands. Generally, it is better to mix the manganese material with a fertilizer and place it in sands.

Another successful method is to spray the soluble manganese sulfate, usually at the rate of 5–10 lb $MnSO_4$/acre once during the growing season. This corrects deficiency symptoms and supplies the crop with a sufficient amount for normal growth.

CONVERSION FACTORS

	Multiply by
Manganese (Mn) to MnO	1.29
MnO to Mn	0.77
Mn to MnO_2	1.58
MnO_2 to Mn	0.63

General References

Stiles, W., "Trace Elements in Plants," 3rd Edition, Cambridge University Press, 1961.

Mitchell, R. L., "Spectrographic determination of trace elements in soils," *J. Soc. Chem. Industry (London) Transactions*, **55**, 267–269.

Mulder, E. G., and Gerretsen, F. C., "Soil Manganese in Relation to Plant Growth," *dv an. Agron.*, **IV**, 222–272 (1952).

McHargue, J. S., *J. Amer. Chem. Soc.*, **44**, 1592 (1922).

"Hunger Signs in Crops," 3rd Edition, David McKay Co., N.Y., 1964.

8

BORON

Modern research has demonstrated the vital importance of the inorganic plant nutrients now collectively known as trace elements or micronutrients. This group, once hardly considered at all in plant nutrition practices, is now receiving major attention from soil scientists and plant physiologists. This is particularly true of the nutrient boron (B) and its role in plant nutrition. Considerable literature has developed concerning boron, much of which is descriptive in nature rather than quantitative in results from fundamental types of research. The disease symptoms of shoots and roots of most of the important agricultural plant species, which occur from deficiencies or excesses of boron in the soil, have been thoroughly described and are generally well-known to most agronomists.

The fact that boron occurs naturally in plants was first discovered by A. Wittstein and F. Apoiger in 1857[1]. It was H. Jay, however, who was able to suggest in 1895[2], after exhaustive analyses of many plants and plant products, that boron might be universally distributed throughout

---

[1] Wittstein, A., and Apoiger, F., "Entdeckung der Borsäure im Pflanzenreiche," *Ann. Chem. Pharm.*, **103**, 362–364 (1857).
[2] Jay, H., "Sur la dispersion de l'acide borique dans la nature," *Compt. Rend. Acad. Sci. (Paris)*, **121**, 896–899 (1895).

the plant kingdom. It is now generally accepted that boron occurs universally in all autotrophic plants, in some of the heterophytes such as the mushrooms, and in various quantities in soils. Also it has been established that plants differ widely in their ability to absorb boron from soils and water and in their requirements of the element.

One of the most comprehensive surveys of the distribution of boron in the soils of America was made in 1942 by R. R. Whetstone, W. O. Robinson, and H. G. Byers, and published as *U.S.D.A. Bu. Plant Industry Technical Bulletin 797*. These authors reported that natural boron toxicity is likely to occur in arid regions where soluble boron salts may accumulate in the soil; that soils formed under normal rainfall contain from 4–88 ppm total boron, and the range of available, soluble boron extends from 0.4–64.8 ppm and averages 17.1 ppm, which represents about 50% of the total. In 300 soil samples examined they detected the presence of boron. Another more recent survey, confined to the state of New Jersey, was conducted by Firman E. Bear and associates in 1944 and published as *N.J. Agr. Expt. Station Bulletin No. 709*. About 350 widely distributed farms were involved, on which test plots were placed to study crop responses. The bulletin has served as a guide for agricultural practices involving boron nutrition of plants in crop production, particularly on soils of the Atlantic Coastal Plain.

The first investigator to produce convincing evidence of the essential nature of boron as a plant nutrient was the French scientist P. Mazé, who published his results in 1914, 1915, and 1919[3]. Although Mazé used only one species, the corn plant, in his investigations, the evidence apparently was so convincing that his conclusions was warranted. His general conclusions were later confirmed by the extensive and thorough investigations carried out in England by K. Warington with the broad bean and later by Warrington and W. E. Brenchley[4].

The work which decisively decided the question of essentiality of boron in plant growth was that published by A. L. Sommer and C. B. Lipman in 1926 and A. L. Sommer in 1927[5]. These investigators, by means of the most painstaking techniques, established definitely the essentiality of boron for a large number of plant species, including

[3] Mazé, P., *Ann. Inst. Pasteur*, **28**, 21–68 (1914); *Compt. Rend. Acad. Sci. (Paris)*, **160**, 211–214 (1915); *Ann. Inst. Pasteur*, **33**, 139–173 (1919).

[4] Warington, K., *Ann. Botany* **37**, 629–672 (1923); Brenchley, W. E., and Warington, K., *Ann. Botany*, **41**, 167–187.

[5] Sommer, A. L., and Lipman, C. B., *Plant Physiol.*, **1**, 231–249 (1926); Sommer, A. L., *Science*, **66**, 482–484 (1927).

corn, sunflower, peas, vetch, barley, buckwheat, dahlia, potato, lettuce, millet, sugar beet, castor bean, sorghum, pumpkin, mustard, and flax. Their conclusion, that the higher green plants require boron in their diet, was accepted as valid by the scientific world.

Boron deficiency was recognized in the field in the southern section of the United States starting in the late 1930's and was often connected with the practice of liming. Soils of this region that are coarse-textured or very sandy and low in organic matter tend to be naturally low or deficient in boron, since such soils favor leaching and are unable to retain much moisture. Wear and Patterson showed[6] that a soil with a high pH reaction (above pH 7) was low in available boron and that heavy liming of sandy soils induced a significant lowering of available boron content.

Boron: its Chemistry

Boron belongs to Group III in the Periodic Table. The other members of this group are aluminum, gallium, indium, and thallium. All three elements except boron are classified as metallic, with boron as a semimetal. The special character of boron derives principally from the relatively small size of the boron atom; it is so small that if a B^{3+} ion were placed in water, it would pull electrons to itself from H—O—H sufficiently to rupture the O—H bond and release H^+ ions; in other words, $B(OH)_3$, and the corresponding oxide, B_2O_3, are acidic[7].

In nature boron is moderately rare and occurs principally as the borates of calcium and sodium, for example, Borax ($Na_2B_4O_7 \cdot 1OH_2$). Pure borax contains 11.36% B or 26.5% B_2O_3. Fertilizer grades are available; one contains 14.30% B or 46% B_2O_3 and a highly concentrated, soluble grade containing 18–20% B or 58–65% B_2O_3. These materials are applied alone or with other fertilizer materials, while the soluble grade may be applied as a dust or in solution as a spray, alone or with insecticides.

At room temperature boron is inert except to strong oxidizing agents, such as fluorine and concentrated nitric acid. When fused with alkaline, oxidizing mixtures, such as $NaOH$ and $NaNO_3$, it forms borates. The only important oxide is boric oxide (B_2O_3), which is

[6] Wear, J. I., and Patterson, R. M., Soil Sci. Soc. Am. Proc., 26, 344–345 (1962).
[7] Sienko, M. J., and Plane, R. A., "Chemistry," McGraw-Hill Book Co., 1957.

acidic, dissolves in water, and forms boric acid (H_3BO_3), an extremely weak acid (this is the reason why it can be used safely as an eyewash).

As elsewhere noted, in the soil boron is found only in combination with oxygen in the form of insoluble tourmaline. The main kinds of available boron present in soils are those derived from marine sediments or plant residues. Accordingly, those boron containing minerals occurring naturally in igneous rocks and sandstones are quite valueless for plants, whereas the boron compounds in marine clays and shales are valuable as readily available forms. As previously mentioned, boron in available forms is readily leached from open type soils and, consequently, highly leached acid soils are generally low in available boron. Boron availability is also diminished by liming podzolic soils[8]. The relative solubility of boron compounds is shown in Table 8.1.

Although boron was established as essential to higher plants only about five decades ago, the fund of knowledge concerning its importance in plant nutrition, which has since accumulated, is very considerable. A great deal is known of the symptoms of deficiency and excess in a wide range of crops, together with ways and means for prevention or control. Its physiological role, however, is still not completely certain. It has been suggested, for example, that boron is involved in the metabolism of protein, in the synthesis of pectin, in maintaining the correct water relations within the plant, in the resynthesis of adenosine triphosphate (ATP), in the translocation of sugar, in the fruiting processes, in the growth of the pollen tube, and in the development of the flowering and fruiting stages. What is definitely certain, however, is that under conditions of boron deficiency, the yield of crops is substantially reduced.

Despite the imperfect knowledge of its fundamental role in the plant's physiology, it has been shown by experiment that only a

TABLE 8.1.   SOLUBILITY IN GRAMS OF ANHYDRIDE PER 100 GRAMS COLD WATER

Boric acid, H_3BO_3	5.15
Potassium tetraborate, $K_2B_4O_7 \cdot 5H_2O$	26.70
Sodium tetraborate, borax, $Na_2B_4O_7 \cdot 10\,H_2O$	1.60
Calcium metaborate, $Ca(BO_2)_2 \cdot 2H_2O$	0.31

Source: Cooper, H. P., Clemson, A. E. S., and Clemson, S. C.

[8] Beeson, K. C., *Soil Sci.*, **60**, 9 (1945).

portion of the boron contained in the plant is available for its use, and even this is not very mobile. This explains why, when this amount is used up, it is necessary to supply a continuous addition to maintain the normal growth process. Failing in this, symptoms of deficiency appear which may show up as stunted or distorted growth, discoloration, and abnormalities in fruit and foliage, failure in reproduction, and death of growing tips followed by death of part or all of the plant. The net result is a reduction in the yield and quality of the crop.

During the years 1953–1956 a series of experiments were carried out in Great Britain to determine the effect of boron deficiency on vegetative growth and the yield of seed of leguminous crops, which are known to require a high amount of boron[9]. The results showed strikingly that boron is primarily needed to maintain the apical growing points and is directly concerned in the process of cell division. It was also established by experimental plants of alfalfa, red clover, and field beans (all legumes) that different plant species require different amounts of boron and that the vegetative yield does not depend directly upon boron supply except over a narrow range (the yield of red clover given by 0.001 ppm boron was as great as that given by higher levels up to 2.5 ppm, and the yield of field beans reached a maximum at 0.05 ppm and was not affected by increases up to 2.5 ppm). The same series of experiments demonstrated that boron deficiency greatly reduced the number of flowers formed.

Species of plants vary widely in their requirements of boron; the rates recommended for alfalfa and beets would injure cereal grains and snap beans. German research into boron as a plant nutrient has been quite extensive, and it is worthwhile to highlight some of their results as reported by Dr. K. Scharrer, the well-known German authority.

W. Oetting[10] found that boron accumulated in the leaves and reproductive organs and less was found in the roots and fruits. The boron of seawater accumulates in the sea weeds, and this explains the high boron content of red algae. Steinbeck[11] reported that the reason potato tubers from Salzburg discolored rapidly when cut was due to boron deficiency. Other experiments on the same soil showed marked response to applied borated fertilizer by potato plants on plots which

[9] Whittington, W. J., "The role of boron in the nutrition of certain legumes," Ph.D. Thesis, University of Nottingham, U.K., 1956.
[10] Oetting, W., Z. Pflanzenernaehr. Dueng. Bodenk., 55, 235 (1951).
[11] Steinbeck, O., Pflanzenernaehr. Dueng. Bodenk., 64, 154 (1954).

had shown deficiency symptoms; the yields of the borated plots were four times higher than the plots not treated with boron, and the starch content of the untreated potatoes was about 1% less.

In order to avoid damage by excessive application of boron, A. Jacob and co-workers[12] proposed that minerals containing boron in the water insoluble form be used in place of water-soluble borax. (In the United States frits are advocated for this purpose). Since the danger of toxicity from excessive dosages is most frequent in the early stages of a plant's growth, the less water-soluble boron should be beneficial. Cook showed[1] that in the mineral colemanite, a boracic calcite, the boron content is as available to plant life as that of borax, even though it is not soluble in water. Jacob found, however, that in the early stage of growth, the amount of boron supplied by colemanite was considerably less than that from borax, but, at a later stage the absorption from colemanite equals that from borax and therefore it may be considered an effective supplier of boron.

The treatment of vineyards with borated fertilizer has become an important practice in Germany and other countries. Boron deficiency has increased in vineyards because of drought and heavy fertilization with nitrogen and lime. Application of borax up to 200 kg/hectare (1.78 cwt/acre) has been effective in preventing boron deficiency. This one application is expected to suffice for several years' requirements.

The search for a suitable indicator plant to warn about an incipient deficiency of boron has led to the use of the turnip plant or sunflower seedlings, grown in boron free nutrient solutions.

Deficiency in Soils and Crops

Boron deficiencies in United States crops have been reported by research workers in each of the 50 states. In a 1962 survey boron was reported as deficient in one or more crop species in 41 states: boron deficient alfalfa in 38 states; beet, in 12; celery, in 10; clover, in 13; crucifers, in 25; and fruit trees, in 21. Deficiencies of boron led all other micronutrient deficiencies. Recent reports on such deficiencies came from areas and crops that previously had been unknown. Boron deficiencies occur more frequently in the more humid regions of a

---

[12] Jacob, A., et al., *Landwirtsch. Forsch.*, **6**, 10 (1954).

country, on highly leached, acid soils, and on organic and alkaline or calcareous soils. Apparently available boron is readily leached from soils[13]. Boron is recommended for responsive crops when the soil contains less than 1 ppm.

Boron occurs naturally in soils as the mineral tourmaline (about 10% B), and although this mineral is the chief source of boron, a considerable portion of the soil's content of boron is held in its organic matter, from which it is gradually released by soil microorganisms. During periods of drought when a soil's microbial life is at low ebb, it is observed that boron deficiencies increase, whereas when soil moisture is adequate, a satisfactory amount is released and supplied to plants. The total content of boron in soils is estimated to vary between 20 and 200 lb/acre plow depth, but this total is not a reliable guide to the adequacy of boron for the needs of the crop growing at any one time because of the modifying chemical, physical, climatic, and biological factors which influence availability.

The inorganic forms in which boron occurs in soils are chiefly borates of calcium, magnesium, and sodium, which result from the slow dissolution of boron containing minerals, mainly tourmaline. Soil organisms compete with crop plants in utilizing such available boron; they transform it into organic forms which, upon their death, are released and oxidized to the various inorganic types. Losses of soluble boron are due mainly to leaching, removal by harvesting crops, and reversion to unavailable, inorganic compounds. The response of crop plants to available soil boron is accordingly influenced by one or more of the following factors: soil acidity, moisture supply, lime content, crop to be grown, concentration of boron, status of organic matter, amount of leaching, and purity of the chemical fertilizer applied. The extent to which one or more of these factors operates on the available boron supply will determine the amount of boron available to the crop. If the amount of effective boron in the soil is increased beyond a limit critical for the planted crop, toxicity symptoms will develop to signal danger.

Deficiency Symptoms (see Figures 8.1, 8.2, 8.3, 8.4, 8.5, 8.6)

Plants deficient in boron exhibit many different symptoms. The number of possible symptoms or phases of deficiency for each crop is

[13] Berger, K. C., "Introductory Soils," MacMillan Co., 1967.

very high, especially when it is considered that plants require such extremely small amounts of this element. Researchers at the Vermont Agricultural Experiment Station in Bulletin 501 (1943) reported seven different symptoms of deficiency of alfalfa on one soil, namely, terminal dieback, rosetting, multiple branching, seed stripping, defective inflorescence, seedling death, and abnormal coloring of foliage. These authors stated that, provided the proper degree of boron deficiency is present, every kind of plant can exhibit at least ten different symptoms. The extent or degree of each symptom depends on the time and stage in the plant's growth cycle at which the deficiency occurs. Some symptoms for each crop have a dollar and cents importance, because they affect yield and quality of the crop, others do not, because their effects are not easily observed.

The experts are well agreed that a deficiency or excess of boron produces characteristic symptoms. During the past 25 years the effects of boron on plant life have been extensively studied and documented, perhaps more than those of all the other trace elements. A number of descriptive terms have been applied to boron deficiency symptoms, among which are brown heart or brown rot of turnips and beets, cracked stem of celery, hollow stem of cauliflower, black spot and canker of garden beets, measles and corky core of apples, dieback and fruit cracking of peaches, monkey face fruit of olives, short heads of barley, and yellows of alfalfa, and browning of cauliflower curds.

In naturally grown plant leaves, the content of boron averages about 70 ppm, although Cook found 283 ppm in the aboveground parts of cowpeas and 152 ppm in radish roots grown in Florida[14] and McHargue reported 154 ppm in hickory leaves[15] (see Table 8.2).

Boron deficient cotton exhibits short internodes and a profusion of branches. Most of the fruit buds abort and few mature bolls are formed.

Boron in Fertilizers

It is necessary to be extremely careful in the formulation of borated fertilizers, due to the toxicity to certain plants when excessive amounts of the element are included in the mixture. The optimal quantity of

[14] Cook, F. C., *J. Agr. Res.*, **5**, 887–890 (1916).
[15] Eaton, F. M., *J. Agr. Res.*, **69**, 237 (1944).

**TABLE 8.2.   RANGE   OF   CONCENTRATION   OF
TRACE   ELEMENTS   AS   REQUIRED
FOR   NORMAL   PLANT   GROWTH**

TRACE ELEMENT	CONCENTRATION PARTS PER MILLION (ppm)
Iron (Fe)	0.5 –5.0
Manganese (Mn)	0.1 –0.5
Boron (B)	0.1 –1.0
Zinc (Zn)	0.02–0.2
Copper (Cu)	0.01–0.05
Molybdenum (Mo)	0.01–0.05

Source: Pillai, K. M., Coimbatore. India.

boron for mixed crops may vary from 5–20 lb borax/acre depending upon the crop; perhaps the inclusion of 10 lb borax/ton of fertilizer for soils in the humid region receiving an application of 1,000 lb fertilizer/ acre may be good practice. Where the fertilizer application is one ton to the acre, it may be advisable to limit the amount of borax in the mixture to 5 lb for most crops. Alfalfa is quite tolerant of borax, and it is usual to add 20 lb of borax/acre for this crop, particularly on soils which have been heavily limed (pH 7.5–8.0).

The more intensive type of farming practices which prevail in most modern farming areas may make it necessary to include some boron in the fertilizer for most crops. It has been observed that many of the legume, root, tuber, and cotton crops respond significantly to borated fertilizers when applied in appropriate dosages.

The most commonly used boron compound in formulating solid fertilizers is borax, whose chemical formula is sodium tetraborate $(Na_2B_4O_7 \cdot 10H_2O)$. It contains 11.34% B and 10 moles of water as water of crystallization. Other commercially available forms are "Tronabor" containing 14.30% B, and a highly soluble borax containing 18–20% B for spray application. Borax is also prepared as fritted boron, a slowly soluble glasslike material containing about 6% B.

Borax cannot be added to finished, acid forming fertilizers containing nitrogen, because the water of crystallization is released causing the fertilizer to become moist. Manufacturers have found that it is satisfactory to add borax early in the processing phase because subsequent drying removes most of the released water. The phosphate-

8.1 Boron deficiency, showing blossom blasting on apple tree. (*Courtesy of National Plant Food Institute*)

potash fertilizer mixtures (PK) do not have the released water problem, and the free acid of the phosphate is neutralized with lime.

The manufacture of commercial fertilizers containing trace elements is introducing many new production, as well as application and promotion problems, to the industry. Recent surveys have revealed a deficiency of various plant nutrient elements in many areas, which apparently had not reported such previously. The industry is

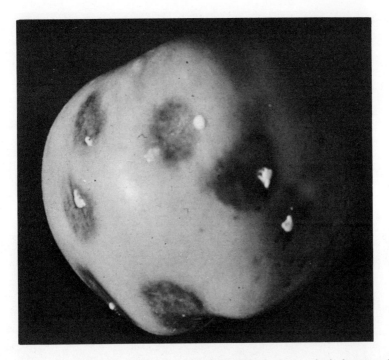

8.2 Immature apple, showing boron deficiency with the ooze which forms as the cork tissue first develops. (*Courtesy of National Plant Food Institute*)

co-operating with state and federal agencies in studying ways of alleviating these deficiencies. But it will take time to develop proper methods of manufacturing fertilizer mixtures with locally required amounts of trace elements, besides first learning how to make, handle, and store such mixtures. To illustrate the manufacturer's problems, a state, let us say Wisconsin, reports numerous areas having an acute shortage of boron, with about 60% of its 2.5 million acres of alfalfa showing a boron deficiency. The authorities recommend top-dressing with 0-13-20 fertilizer, 10% of the fertilizer to be borax. The manufacturer faces the serious problem of where to find equipment for properly incorporating the borax uniformly in the fertilizer. He has to use separate bins for the borated mixes, which ordinarily are made to order in small tonnage lots for individual grower customers. Usually such a borated mix cannot be applied to any other crop besides alfalfa. This situation regarding the tailor-made, trace element fertilizer mix

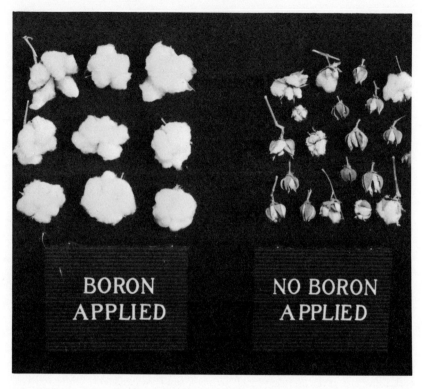

8.3 Boron deficiency in cotton. (*Courtesy of the American Potash Institute*)

is common, since even in one locale the deficiency differs in degree, and if an excess is used on a crop with a small tolerance limit or in a soil where the deficiency is minor, serious injury to the crops may occur.

Another drawback from the manufacturer's viewpoint is that by adding the required trace elements to the fertilizer, which is formulated to grade, that grade becomes diluted, and the state regulatory agency requires that the manufacturer procure another license. Fertilizer producers find it hard to adopt any specific quantity of a trace element for any specified grade of fertilizer, for the reason that farmers apply the major fertilizer grade at different rates for the same crop and also the degree of deficiency varies from farm to farm. Soil and seasonal climatic conditions will also influence the response of crops to applications of one or more micronutrients. Hence, the manufacturer prefers to produce the tailor-made type of trace element fortified

TABLE 8.3. BORON CONTENT OF SPECIFIED PLANTS

CROP	B CONTENT OF OVEN-DRIED MATERIAL ppm
Alfalfa	25–40
Barley	2– 4
Cabbage	30–40
Corn (whole plant)	4– 6
Corn (leaves)	10–25
Peas	15–25
Potato	10–20
Red beet	35–60
Soybean	20–30
Tobacco	20–30
Tomato	15–25
Wheat	3– 5

Source: Berger, K. C., loc. cit.

fertilizer, and experience proves such practice is generally also more economical for the farmer.

It is easy to understand the difficulty of obtaining proper distribution of the small quantities of boron or other trace element required on an acre of ground if made separately. The problem might be solved by incorporating the element uniformly in a fertilizer mixture. When most fertilizers were pulverant, the problem was simpler, but with today's granular type fertilizers predominating, the problem is ever so much greater—the powdered trace element carrier readily segregates from the granular mass. The ideal method would be to incorporate the trace element material in the granule of the NPK fertilizer. For the reasons previously cited, such a procedure is not possible, owing to the numerous variations in amounts and rates of application. TVA* researchers have proposed a method by which the granules are uniformly coated with the micronutrient carrier. The proposed method is still in the experimental stage. It does have the advantage of preventing segregation but also the disadvantages of using a dusty, hard-to-handle pulverant carrier, an oily sticker or binder and additional equipment. A practical solution is to supply

* TVA = Tennessee Valley Authority, Fertilizer Research, Muscle Shoals, Alabama.

8.4 Boron deficiency, showing celery plant badly affected with cracked stem disease. (*Courtesy of Michigan State University*)

borax or other trace element in granular form, the size and density more or less the same as those of the regular granular fertilizer. Such granules would not segregate and the mixing operation would be fairly simple. Granular (− 16 + 48 mesh) boron salts are now available.

Official Recommendations on Boron Use

PEANUTS

Georgia

Through the years concealed damage (hollow heart damage), due to deficiency of the minor element boron, has occurred most often on the lighter soil types in dry years (see Fig. 8.5). Using ½ lb of actual boron will cost approximately 50 ¢/acre and could well be used as an insurance practice on all light soils. Farmers are urged to consider using this measure if peanuts are planted on light textured soils.

Boron may be effectively used as a foliar application in either dust or spray form. If one prefers to spray, Solubor* is available and is compatible with most insecticides recommended for peanut insect control. Dust or sprayable forms of boron should not be applied any later than the early blooming stage for best results. Peanut foliage should also be

---

* Registered trade name.

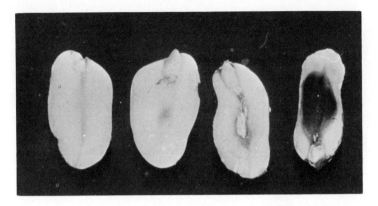

8.5 Boron deficiency in peanuts, showing increasing severity of hollow heart, left to right. (*Courtesy of Virginia Polytech. Inst. Ext. Service*)

dry when making foliar applications, to reduce the possibility of burn to peanut foliage (McGill J. F., Extension Agronomist—Peanuts, Cooperative Extension Service, University of Georgia).

Oklahoma

Fertilizers have influenced the quality of peanuts. The farmer receives payment for peanuts based on the weight of sound mature kernels delivered. Elimination of "pops" internal damage and shrivelled kernels represent quality improvement. Four fertilizer elements are reported in the literature as affecting peanut quality in this sense; calcium, potassium, boron, and copper.

Boron was shown to have a significant beneficial effect on two varieties of peanuts by H. C. Harris (*Soil Sci.*, **84**, 233–242 (1957)).

Data obtained by Tucker and Chrudimsky of Oklahoma State University also show a significant improvement in crop quality where boron deficiency causing "internal damage" has been noted.

The best yielding crop, producing the most sound mature kernels (SMK), was grown after fertilizer treatment containing 0.8 lb of boron/ acre. The critical nature of boron application was evidenced by the decrease in yield when 1.6 lb of boron was added to the fertilizer.

Southeast

Alabama, Georgia, North Carolina, and Virginia recommend 0.5 lb/acre in one of the following methods:

(1) Borated fertilizer.

8.6 Boron deficiency in sugar beets, shown by "crown rot" and "canker". (*Courtesy of Mich. State College*)

(2) Dusted or sprayed on foliage.
(3) Incorporated with pesticide and dusted or sprayed on foliage.
(4) Borated gypsum.

Care should be taken that the rate of $\frac{1}{2}$ lb of actual boron (B)/acre not be exceeded. Excessive amounts may reduce yields. Dusting should not overlap, and preferably should be done when the foliage is dry.

Florida

Comparisons of calcium and boron deficiency and normal peanuts were made from plants grown to maturity on soils supplying inadequate and adequate (with fertilization) levels of these nutrients. The characteristics of boron and calcium deficiencies were distinctive and strikingly different. Boron deficiency seemed to affect all parts of the plant. It changed foliage characteristics, flowering pattern, and usually increased shoot and root yields. However, the deficiency decreased fruit yield and caused poor quality of fruit. In contrast, calcium deficiency seemed to affect only fruit yield and quality. Without calcium the yield usually was lower and the quality of seed poorer.

Both calcium and boron applications usually considerably increased the yield of mature pods. One compartment pods were numerous when boron was not applied. The percentage of plump normal seed was greatly increased by the application of these two elements. Thus, both calcium and boron applications markedly increased pod yields and improved the grade of seed. Boron appeared to have more effect in this respect (H. C. Harris, *Agron. J*, **58**, 575–578 (1966)).

SOYBEANS

Arkansas

Several years ago research at the Arkansas station showed that some eastern Arkansas soils used for cotton are low in available boron and that adding boron, either to the soil or to the cotton foliage, increased yields (*Arkansas Farm Research*, Vol. **12**, No. 4, and Vol. **13**, No. 2). As a result of this and other research work, boron has been applied to about 400,000 acres of Arkansas cotton soils, especially the slightly acid to alkaline soils, where the soil boron is less available than in more acid soils.

It has been known for some time that germinating seeds of the large seeded legume species are especially sensitive to boron salts applied close to the row. Soybeans are sometimes planted following a cotton stand failure. There has been some concern about possible damage to soybeans planted in soil that had been treated with boron. Two field experiments conducted in Arkansas provide information on this problem.

In 1963 a soybean experiment was established in Prairie County on an acid Henry–Calloway silt loam that had been cleared of regrowth timber two years earlier. This infertile soil had a pH of 4.9, contained 1.0% organic matter, and 600 lb exchangeable calcium/acre.

Boron, as a solution of Solubor (a patented, highly soluble boron fertilizer containing 20.5% boron), was applied preemerge in 10-in. bands over the row. The rates of boron applied and the resulting soybean yields are given in Table 8.4. It can be seen that boron neither increased nor decreased yields.

The second experiment was located at the Southeast Branch Station in Desha County in 1965. This experiment had been established

TABLE 8.4. EFFECT ON SOYBEAN YIELDS OF BORON, APPLIED PREEMERGE ON AN ACID, LOESSIAL TERRACE SOIL, 1963

RATE PER ACRE AS		YIELD,
BORON	SOLUBOR	bu/acre
0	0	22.6
0.5	2.5	24.5
1.0	5.0	22.7
1.5	7.5	23.2
2.0	10.0	22.1
3.0	15.0	20.6
6.0	30.0	21.3

in the spring of 1964 as a factorial study of limestone and boron rates on cotton. In 1965, after cotton had been planted twice without obtaining a stand, Lee soybeans were seeded on June 16. The soil was a Gallion sandy loam. Its pH and exchangeable calcium content are shown in Table 8.5. The borated plots had received 0.5 lb boron/acre each year. The 1965 application was made as a solution of Solubor in a 12-in. band over the row on May 21.

As can be seen from the soybean yields given in Table 8.5, limestone significantly increased yields by from 2–3.5 bushels/acre. Boron also increased yields, by 1.3 bushels/acre. This was significant at the 11.3% level, meaning that in about 88 chances in 100, the yield increase from boron actually was caused by the boron application. The limestone-boron interaction was not significant.

This work indicates that if soybeans are planted following cotton, on a soil that has been fertilized with recommended rates of boron, the boron will not adversely affect soybean growth or yield. Boron, of course, is not recommended for soybeans (by Thompson, Lyell, and Hardy, Glen, *Arkansas Farm Research* (March–April, 1967)).

General Considerations

Table 8.6 lists boron deficiencies in major crops, together with general symptoms and the suggested treatment with boron. Of course, such a list must be considered as being of a general nature, since it is supposed

TABLE 8.5.   EFFECTS OF LIMESTONE APPLIED IN APRIL, 1964, ON SOIL pH AND EXCHANGEABLE CALCIUM, AND OF THIS LIMESTONE AND BORON ON SOYBEAN YIELDS IN 1965

MEASURE	LIMESTONE, TONS PER ACRE				AVERAGE
	0	1	2	4	
Soil pH	6.00	6.57	6.86	7.00	—
lb of calcium/acre	1892	2350	2765	3163	—
	SOYBEAN YIELD, BUSHELS PER ACRE				
No boron	28.0	29.2	29.8	30.5	29.4
0.5 lb boron/acre*	27.8	30.6	32.4	32.1	30.7
Weighted average†	27.9b	30.1a	31.5a	31.6a	—

* These plots also received 0.5 lb boron/acre in 1964.
† Averages followed by different letters differ significantly at the 5% level.

to cover the entire country. Many states have issued similar lists which apply to their local areas and crops. Several such local and regional releases follow: cotton for the seven Southeastern States (Alabama, Arkansas, Georgia, Mississippi, North Carolina, South Carolina, and Tennessee); peanuts for Georgia and Oklahoma; Soybeans for Arkansas; alfalfa for Wisconsin; and tree fruits for Washington state.

UNIFORM BORON RECOMMENDATION FOR COTTON NEEDED IN THE SOUTH-
EASTERN U.S.

At present seven of the Southeastern states recommend boron for cotton fertilization on all or certain soil areas. Since each of these states has slightly different recommended boron applications for cotton, but all averaging about ½ lb of actual boron/acre, it would seem that the maximum effectiveness of this recommendation might be increased by a uniform recommendation for all states involved.

It would greatly assist the fertilizer manufacturers operating in the Southeast and across state lines if a uniform recommendation for boron could be adopted. Some of the particular advantages that would result are as follows:

(1) Fertilizer manufacturers could add boron during manufacture of the fertilizer, and thus obtain a more uniform mixture and avoid possible segregation of the boron. This would also eliminate the common practice now of adding boron on demand at the time of shipment.

TABLE 8.6. BORON DEFICIENCY IN MAJOR CROPS*

CROP	SYMPTOMS	SUGGESTED BORON (B) TREATMENT: lb/acre†
	FIELD CROPS	
Alfalfa	Death of terminal bud, rosetting, yellow top, little flowering, and poor seed set.	1–3 lb
Beets, sugar	Yellowing or drying of leaves, cracking of leaf midrib, brown discoloration of internal tissue, rotting of crown.	1–2.5 lb
Clover: crimson, red, white	Poor stands, growth and color, reduced flowering and seed set.	1–1.5 lb
Cotton	Excessive shedding of squares and young bolls, rupture at base of squares and young bolls, dark colored exudate from ruptures, dark internal discoloration at base of boll, half opened bolls, green leaves until frost.	0.2–0.5 lb in drill; 0.5–1.0 lb broadcast or 0.5 lb in foliar sprays.
Corn	Short, bent cobs, barren ears, poor kernel development.	0.5 lb.
Peanuts	Dark, hollow area in center of peanut called "hollow heart."	0.3–0.5 lb
Tobacco	Leaf puckering, deformed buds.	0.25–0.5 lb
	VEGETABLE CROPS	
Beets, red	Black spot, internal breakdown.	1–2 lb
Broccoli	Brown, dry areas in heads.	1–2 lb
Cabbage	Internal breakdown of stem.	1–2 lb
Carrots	Death of growing tip, yellow leaf margin.	1–2 lb
Cauliflower	Deformed foliage and brown curd.	1–2 lb
Celery	Cracked stem, mottling of bud leaves.	1–2 lb

TABLE 8.6 (continued)

CROP	SYMPTOMS	SUGGESTED BORON (B) TREATMENT: lb/acre†
	VEGETABLE CROPS (continued)	
Lettuce	Death of terminal bud, brittle cup shaped leaves.	1–2 lb
Radish	Pale chlorotic leaves, darkening of root tissue.	1–2 lb
Sweet potatoes	Internal brown spot, corkiness.	0.5–1 lb
Tomatoes	Dwarf plants, orange-yellow foliage.	1–1.5 lb
Turnips	Internal browning ("brown heart").	1–2 lb
	FRUIT AND NUT CROPS	
Apples	Rosette and dieback of shoots, corky core of fruit, drought spot.	1–2 oz./ mature tree.
Citrus: Florida	Death of young shoots, hard, misshapen fruit, low juice and sugar content.	Fertilizers: 0.1–0.3% B_2O_3. Nutritional sprays: $\frac{1}{2}$ lb Borax Eq. per 100 gal.
Grapes	Dieback of shoots, shortened internodes, poor fruit set, "shot berries."	1 oz borax of FB-46 per vine. $\frac{1}{2}$ oz FB-65 or Solubor per vine.
Pears	Blossom blast, dwarf terminal leaves.	1–2 oz/per mature tree or 5–8 lb per acre Solubor in sprays.

TABLE 8.6 (continued)

CROPS	SYMPTOMS	SUGGESTED BORON (B) TREATMENT: lb/acre†
	FRUIT AND NUT CROPS (continued)	
Prunes	Excessive multiple branching in tree tops, "Bushy Branch"; marked reduction in fruit set, brown sunken areas in fruit.	1 lb B per acre in spray in early summer, or $\frac{1}{2}$ lb borax per tree.
Walnuts	Leaf "scorch," irregular dark brown spots between vines; "snakehead" on terminal branches; low yield of nuts.	$\frac{1}{2}$ lb B per 18–25 yr old trees broadcast in spring—decrease to $\frac{1}{3}$ lb B for 12 yr old trees.

* BORON-O-GRAM NS 89 1966: U.S. Borax and Chemical Corporation, Auburn, Alabama.
† Note that except for fruit and nut crops the amount of boron is listed as element in keeping with Model Fertilizer Bill. For specific recommendations local agricultural advisors should be consulted.

(2) Recommendations of boron could be more uniformly followed and applied to soils by farmers.

(3) It would save cost in manufacture, since relatively large batches of fertilizer could be processed with the proper amount of boron included along with the other raw materials in manufacture.

Perhaps it would be necessary to have one standard addition of boron for regular analysis fertilizers such as 6-12-12 and another rate of addition for high analysis fertilizers such as 8-24-24. For example, if it is assumed that approximately $\frac{1}{3}$ ton of 6-12-12 fertilizer is applied/ acre, then a ton of fertilizer should contain $1\frac{1}{2}$ lb of actual B/ton. An 8-24-24 fertilizer at the rate of $\frac{1}{6}$ ton/acre should contain 3 lb of actual B/ton of fertilizer. Slight variations of application would be of little or no concern, since the range in boron recommendations varies from about 0.3–0.6 lb of B/acre.

**TABLE 8.7.   COTTON: RECOMMENDED RATES AND METHODS OF
APPLICATION WHEN APPLIED IN FERTILIZERS**

STATE	SOILS	BORON, LB OF B PER ACRE	METHODS OF APPLICATION
Alabama	All.	0.3–0.5	Mixed fertilizer at planting.
		0.5–1.0	Borated fertilizer broadcast.
Arkansas	Sandy loams and silt loams B deficient soils.	0.2–0.4	Band in mixed fertilizer or N soln.
		0.5–1.0	Soil and foliar.
Georgia	All soils above pH 5.6.	0.5	Soil and foliar.
Mississippi	Hill section and delta foothills.	0.3–0.5	Borated fertilizers.
North Carolina	pH above 6.5 and deep sands.	0.5	Borated fertilizer.
South Carolina	All.	0.3–1.0	Borated fertilizer, preferably with frit, or side-
		0.6	dress with fertilizer.
Tennessee	pH above 6.0 and where lime is applied.	0.5	Borated fertilizers.

It is believed that the above would be of mutual benefit to the cotton farmers as well as to the fertilizer industry.

The summary in Table 8.7 of boron recommendations for cotton in the Southeast is taken from the official releases for 1966 from the seven southeastern states from North Carolina to Arkansas.

Boron fertilizers may be applied in several ways. At the present time no one method is considered to be the most effective. Therefore, a person may select the method that is most economical and convenient for him. A couple of possibilities are either to use a mixed fertilizer containing boron or to include a soluble form of boron in several insect sprayings. The recommended rate for a season in only $\frac{1}{2}$ lb B/acre. Care should be taken to use as nearly as possible this amount. Amounts much in excess may cause damage to plant growth.

If the boron is added to a mixed fertilizer, the amount of total fertilizer needed to supply $\frac{1}{2}$ lb B/acre should be used.

If boron is to be included in the spray program, a split application is suggested. One can use $\frac{1}{4}$ lb at early bloom and $\frac{1}{4}$ lb 2 weeks later as one system. Another would be to use 0.15 lb each time mentioned above and repeat this later if a drought period occurs.

When boron is to be applied as a spray, a highly water soluble form should be used (F. R. Cox and J. V. Baird, N. C. State Univ. (Boron-o-gram NS96).)

ALFALFA

Wisconsin

A borated fertilizer is recommended for alfalfa when the soil test for boron is below 3.0 lb/acre. Most Wisconsin soils do test less than 3.0 lb/acre.

When boron is needed, 2–3 lb/acre of actual boron (B) should be applied once in the rotation. The ideal borated fertilizer application is as a top-dressing after the first cutting of alfalfa is removed the first hay year. On sandy soils where there is a possibility of leaching, 0.5–1.0 lb/acre of actual boron (B) should be applied every hay year.

The most common boron carriers are fertilizer borate containing 14% boron (46% B_2O_3) and Solubor containing 20% boron (66% B_2O_3). Table 8.8 shows how much boron carrier to add to a ton of fertilizer.

ORCHARDS

Northeast

Shortage of boron in New England and New Jersey soils is so general that all apple and pear orchards in those areas should receive regular application of this nutrient. Many New York and Pennsylvania orchards, particularly second and third generation orchards, are also running into boron shortages, and should receive maintenance applications. The application of 1.5–2.5 lb of boron/acre every third year proves satisfactory. This can be obtained from $\frac{1}{2}$–$\frac{3}{4}$ lb of high grade borate

(14.3% boron) or $\frac{1}{3}$–$\frac{1}{2}$ lb concentrated borate (20.2% boron) per tree every three years.

A shortage of boron can be observed as water-soaked spots occurring in the bark near the growing tips, thus causing the death of the leaves beyond the girdle. The bark eventually gets rough, cracks, and has corklike patches. The fruit may show sunken, corky areas near the skin and core and will tend to fall earlier than normal (see Figures 8.1, 8.2).

Solubor is to be used for maintenance applications and emergency corrective treatments of boron deficiency.

For maintenance use Solubor at the rate of 4–5 lb/acre annually. In dilute application use $\frac{1}{2}$ lb/100 gallons water in the first and second cover. It may be used in a concentrated 2X spray to supply the required boron in two applications.

For corrective treatment the yearly maintenance application of Solubor is doubled. Apply as soon as deficiency is noted or suspected. Do not exceed the concentration recommended above. Apply in more sprays to obtain the required amount.

Caution—Solubor is not compatible with oils. It is compatible with most other present-day spray materials (Ext. from 1967 "Agway Complete Crop Programs for Fruit Growers").

TABLE 8.8.   BORON FERTILIZER GUIDE FOR ALFALFA AT VARIOUS BORON SOIL TEST LEVELS AND FERTILIZER APPLICATION RATES

BORON TEST LESS THAN (lb/acre)	BORON NEEDED PER ROTATION (lb/acre)*	FERTILIZING RATE (lb/acre)	LB OF BORON CARRIER PER TON OF FERTILIZER* (14% B)	20% B)
1.5	3	200	210	150
1.5	3	300	140	100
1.5	3	450	95	65
3.0	2	200	140	100
3.0	2	300	96	65
3.0	2	450	65	45

* On sandy soils, apply one-third ($\frac{1}{3}$) these rates each hay year.
   Do not use a borated fertilizer in the grain drill or corn planter as it may cause poor germination (C. C. Olson and H. B. Pionke, Univ. of Wis. (Boron-o-gram NS 96)).

TREE FRUITS

Washington

Fertilizer recommendations for the White-Salmon-Underwood area non-irrigated tree fruits are nitrogen and boron. To date no evidence has been obtained to show that phosphorus or potassium are needed for tree fruits, but they may be needed for proper cover crop growth in this area.

The boron (soil) test should be made occasionally in orchards, where boron is being used as a plant nutrient to determine if adequate or excessive boron is being applied. The test has proved helpful in diagnosing abnormal growth conditions. Values less than 0.5 ppm are low. Values in the range 0.5–1.0 ppm B indicate the range where tree fruits make normal growth. Values above 1.0 ppm are high and values above 2.0 ppm indicate possible toxicity. Dry soil conditions during much of the growing season reduce boron uptake; therefore, soil applied boron has not been very satisfactory.

Foliar applied boron gives good results. Boron containing materials formulated for spray purposes and known as Solubor and Boro-spray, each 20% B, can be used* at the rate of one lb per 100 gallons water; make one application after harvest before the leaves fall or in the spring at the pink stage. The fall application is most effective in helping fruit set. The spray may need to be repeated annually. Do not apply more than twice in any one year without having the soil checked for its boron content. Boron in amounts only slightly above adequacy may be toxic (from FR-32 Cooperative Extension Service, Washington State University (September, 1965).)

---

* Caution: Do not use use Polybor-chlorate (a herbicide) as a spray. It is harmful to trees, even in dilute solution.

9

ZINC

Zinc is a nutrient element essential to plants and animals. The extensive use of zinc salts as plant nutrients was begun about 1931, although references to such use of zinc had appeared in the literature about 1900[1]. Recognition of the essentiality of zinc occurred in 1928 following the laboratory confirmation, by Sommer and Lipman[2] in California, of a zinc requirement for buckwheat, bean, and barley plants. In 1932, also in California, Chandler, Hoagland, and Hibbard[3] discovered that a long recognized "little leaf" disease of fruit trees was actually a zinc deficiency symptom. Simultaneously, Alben, Cole, and Lewis[4] concluded that "pecan rosette" in the Southeastern States was also caused by zinc deficiency. It was the research of Sommer and Lipman, however, which established zinc as an essential nutrient, and the results were widely accepted.

The first widespread commercial use of zinc salts in the field was on several thousand acres of tung trees in Florida in 1932–33[5]. During

[1] Javallier, M., *Intern. Congr. Appl. Chem.*, **15**, 145–146 (1912).
[2] Sommer, A. L., and Lipman, C. B., *Plant Physiol.*, **1**, 231–249 (1926).
[3] Chandler, W. H., Hoagland, D. R., and Hibbert, P. L., *Proc. Am. Soc. Hort. Sci.*, **29**, 255–263 (1931).
[4] Alben, A. O., Cole, J. R., and Lewis, R. D., *Phytopathology*, **22**, 979–981 (1932).
[5] Camp, A. F., *Soil Sci.*, **60**, 157–164 (1945).

1931–1936 the use of zinc became established as a valuable fertilizer and spray material. During this period zinc deficiency symptoms were definitely described for deciduous fruit trees, pecans, corn, tung, and other crops. By 1936 zinc was in commercial use on citrus in Florida and California and on tung in Florida.

Like iron, zinc is present in nearly all soils in minute quantities, which, theoretically, should be enough to satisfy the needs of ordinary crops. Certain soil conditions, however, may reduce its availability as a plant nutrient; it becomes less available as the pH increases and becomes critical around pH 5.5–6.5; in clayey, acid soils it tends to combine with organic matter, and on open, sandy soils its deficiency is worsened by continued cropping without adequate replenishment.

Chemical and Physical Characteristics

In the Periodic Table zinc is an element in the subgroup consisting of cadmium and mercury in addition to itself. Its atomic number is 30 and it has a valence of 2, that is, it has two electrons in its outermost shell.

Zinc occurs in nature at about one hundred times the abundance of copper and principally as the mineral sphalerite (ZnS), which is also called zinc blende. Metallic zinc is obtained by roasting sphalerite in air, which converts it to zinc oxide. Finely divided carbon is employed to reduce the oxide. The chemical reactions may be represented as follows:

$$2 ZnS(\text{solid}) + 3O_2(\text{gas}) \rightarrow 2 ZnO(\text{solid}) + 2 SO_2(\text{gas}) \tag{1}$$

$$ZnO + C \rightarrow Zn + CO(\text{gas}) \tag{2}$$

Zinc exhibits only a $+2$ oxidation state in all its compounds. The zinc ion (Zn^{2+}) hydrolyzes in aqueous solutions to give slightly acid solutions:

$$Zn^{2+} + H_2O \leftrightarrows Zn(OH)^+ + H^+ \tag{3}$$

white zinc (hydroxide) $(Zn(OH)_2)$ is precipitated when a base is added to solutions of zinc salts. This hydroxide is amphoteric and, when more base is added, it dissolves to give zincate ion, which is represented variously as: $Zn(OH)_3^-$, $Zn(OH)_4^{2-}$, $HZnO_2^-$ or ZnO_2^{2-}.

Zinc has a strong tendency to form stable, complex ions; for example, $Zn(OH)_2$ easily dissolves in aqueous ammonia to form a

TABLE 9.1   SOLUBILITY OF ZINC COMPOUNDS IN GRAMS OF THE ANHYDRIDE FORM PER 100 GRAMS COLD WATER. TEMPERATURE ABOVE 20°C

COMPOUND	SOLUBILITY
Zinc carbonate	0.001
Zinc hydroxide	0.26×10^{-8}
Zinc phosphate	insoluble
Zinc sulfate	86.5
Zinc chloride	432.0
Zinc nitrate	327.3

Source: Cooper, H.P., *Com. Fertilizer* (January, 1947).

zinc-ammonium complex $(Zn(NH_3)_4{}^{2+})$. The hydroxide also dissolves in cyanide solutions, because it can form a zinc-cyanide complex $(Zn(CN)_4{}^{2-})$.

Solubility

The solubility of zinc nutrient carriers varies, as is shown in Table 9.1.

Sources for Fertilizer Uses

Although zinc sulfate is widely used as a source of zinc for fertilizer and spray purposes, several other compounds are also employed. Those most frequently mentioned are the following:

Zinc sulfate, heptahydroxide	$ZnSO_4 \cdot 7H_2O$	23.39% Zn
Zinc sulfate, monohydroxide	$ZnSO_4 \cdot H_2O$	35% Zn on an 89% purity
Zinc chloride	$ZnCl_2$	45% Zn in pure salt
Zinc oxide	ZnO, commercial grade	67–80% Zn
Zinc ammonium phosphate	$Zn(NH_4)PO_4$	36.76% Zn
Zinc oxide-sulfate	$ZnO–ZnSO_4$	about 55% Zn
Zinc chelate (sequestrene) (slowly soluble, for soil or spray)	Zn EDTA	about 6–7% Zn

Liquid zinc		about 1% Zn
Fritted (glass)	FTE	about 4% Zn
(may contain also other		
micronutrients, fritted)		
Sphalerite	ZnS powdered	about 60% Zn, 30% S
Rayplex Zn	—	—
Versenol Zn	—	—
NuZn	—	52% Zn

Zinc in Plants

Zinc is present in all plant tissues. Plant analyses indicate that zinc accumulates in different parts of a plant in the following order, from most to least: root, stem, leaves, fruits. Beeson[6] lists the range in zinc content of some important crops as from 2–3 ppm, dry basis, in the edible portion of certain fruits (apricots, peaches, prunes) to as high as 300 ppm or more in cotton seed and Kentucky Bluegrass seeds; 21 ppm for barley grain, 20 ppm for corn (maize), 22 ppm for oats grain, 70 ppm for wheat grain, 35 ppm for field beans, 83 ppm for oat straw, 17 ppm for wheat straw, 24 ppm for cabbage leaves, 12 ppm for potato tubers, and 18 ppm for turnip roots—all determinations on dry matter basis. Mitchell[7] reported the zinc content of forage in Northeast Scotland as from 5–40 ppm.

Moghe[8] reported on the average zinc content of certain crops grown in India as follows: vegetables, 28.2 ppm; pulses, 34.8 ppm; cereals, 27.8 ppm; fruits, 36.6 ppm; cotton and sugar cane, 36 ppm; grasses, 18.5 ppm.

Research in India[9] on the presence of zinc in the banana plant showed a highly significant effect on total root extension, length of stem, fresh and dry weight of leaves, the number of leaves at 60 and 180 day stages, when zinc was applied as a foliar spray at the concentration of 4 ppm. The effects were more noticeable on the growth of leaves, and other aboveground tissues as well as in the greater extension of the root system.

[6] Beeson, K. C., *U.S.D.A. Miscell. Pub.*, No. 369 (1941).
[7] Mitchell, R. L., *Proc. Nutr. Soc.*, **1**, 183–189 (1944).
[8] Moghe, V. B., *Fertilizer News India*, (Oct. 1965).
[9] Srivastava, R. P., *Fertilizer News India*, 26–32 (June, 1964).

Function of Zn in Plants

Zinc catalyzes the processes of oxidation in plant cells and is vital for the transformation of carbohydrates; regulates the consumption of sugar; increases the source of energy for the production of chlorophyll; aids in the formation of auxins, the growth-promoting compounds; promotes the absorption of water and in so doing prevents stunting. It acts as a component of the enzyme carbonic anhydrase, which is a catalyst that serves to decompose carbonic acid to CO_2 and H_2O and is necessary for the formation of the amino acid tryptophane, itself involved in the elaboration of indoleacetic acid hormone[10].

The zinc content of various plants varies widely. According to Thatcher[11] zinc and copper may be considered a pair of coordinated catalysts in oxidation-reduction phenomena and are particularly concerned in reactions involving the transfer of hydrogen. Since zinc does not change its valence values, it probably participates indirectly in oxidation reactions. Although zinc is required by all plants for normal growth, the amount needed is about one-one hundredth as much as the amount of phosphorus[12].

Considerable evidence exists that serious zinc deficiencies occur in all the major fruit- and nut-producing areas of the Western states. Results of research in Washington (state) indicate that for several crops grown in a normal soil, the zinc content assayed was from 11–32 ppm with an average of about 17 ppm, whereas in areas of zinc deficiency the crop content varied from 9–22 ppm with an average of about 15 ppm.

The information in Table 9.2 is interesting and worthwhile as indicative of the range of zinc values in the specified crops.

The data in Table 9.2 are to be considered fairly reliable standards according to current knowledge. The "normal" levels are judged adequate for high yields, while the "low" and "high" levels apply to problems of imbalance, that is, too little or too much, respectively. In cases which come under the "deficiency" and "excess" levels, crops may suffer serious hurt in both growth and yield effects. "Toxic" levels need no explanation. An intelligent interpretation of the data for any one crop, however, requires the evaluation of other growth factors and definitely of the interaction of all the other plant nutrients

[10] Wear, J. I., and Hagler, T. B., *Plant Food Review* (Spring, 1968).
[11] Thatcher, R. W., *Science*, **79**, 463–466 (1934).
[12] Sprague, H. B., "Hunger Signs in Crops," 3rd Ed., New York, D. McKay Co. (1964).

TABLE 9.2 ZINC CONTENT OF VARIOUS CROP PLANTS

CROP PLANT	PLANT ANALYSIS DATA IN PARTS PER MILLION				
	DEFICIENT	LOW	HIGH	NORMAL	TOXIC
Corn leaves (vegetative stage)	0–10	11–20	21–70	71–150	150
Soybean leaves (vegetative stage)	0–10	11–20	21–70	71–150	150
Wheat, barley, oats (3″–12″ growth)	0–10	11–20	21–40	41–150	150 (excess)
Cotton (vegetative stage)	—	—	20–30	—	—
Tobacco (vegetative stage)	—	0–20	21+	—	—
Sugar beets (vegetative stage)	0–10	11–20	21–70	70+	—
Potatoes	—	0–16	17–40	30+	—
Alfalfa tops	0–8	—	—	9–14	—
Grass (vegetative stage)	—	—	15–80	—	—
Bean (leaves)	0–20	—	—	—	—
Tomatoes (leaves)	0–10	11–20	21–120	121+	—
Citrus (leaves) Florida	0–15	16–25	26–80	81–200	200 (excess)
Tung (leaves)	0–10	11–26	—	—	—
Apples (leaves)	0–15	16–20	21–50	51+	—
Peaches (leaves)	0–16	17–20	21–50	51+	—
Pears (leaves)	0–10	11–16	17–40	41+	—
Grapes (petioles)	—	0–30	31–50	51+	—

Source: M. C. Sparr, et al., *Solutions* (January–February, 1968).

involved. For example, when a low level of one element occurs simultaneously with a general low level of all the essential nutrient elements, yields will be limited moderately; when that same low level, however, is present, and all the other nutrient elements are at a generally high level, then the effect will be a serious limitation on growth and yield.

Visual symptoms of a deficiency or excess of any one nutrient are not always exhibited under most field conditions. Other factors

9.1 Zinc deficient orange leaves. Note small leaves. (*Courtesy of Fla. Agric. Expt. Station*)

such as soil pH, moisture levels, temperature, diseases, and pest in-
festations, may often bring about similar modifications in the foliage
or plant. To pin down such symptoms to one cause is just impossible.
Recourse to plant tissue analysis may help to establish the level of

several nutrients, but the situation is, only too frequently, too complex for a simple solution. Hence, the data in Table 9.2 are necessarily oversimplified, and their limitations must be kept in mind whenever reference is made to them.

Zinc in Soil

Information concerning the soil factors which control the availability of zinc to plants is not too precise. Powers and Pang[13] showed that it is lower in soils having a basic reaction than in acid or neutral soils. McMurtrey and Robinson[14] report that soil analysis data show a general range of from 2–50 ppm with the lowest amount in sandy soils. Although zinc occurs in almost all soils, the amount present is relatively small, the range being from about 20–500 lb Zn/acre, plow depth (2,000,000 lb soil). Plant residues add small amounts of zinc to the soil as they decompose, and this is given as one reason for the presence of more zinc in the upper surface layer of the soil than in the subsoil. Clay minerals, pulverized limestone particles, organic matter, and magnesium ions are believed to be able to adsorb amounts of zinc sufficient to create a deficiency of available zinc. Some of the available zinc is adsorbed as a cation on the colloidal complex of the soil. Actually, however, not much is known about the reactions and interactions of zinc and its compounds in the soil. The zinc ion is not labile and apparently remains in the top soil layer, where it may have accumulated during the years when the soil was cultivated, fertilized, and enriched with organic matter. Tests in North Dakota and neighboring areas in the northern Great Plains indicate that the top soil to plow depth contains, on average, about two-thirds of the total available zinc on the top four feet analyzed. Land that has been levelled for gravity irrigation will generally be deficient in zinc in the exposed subsurface. An application of zinc at the rate of about 15 lb Zn/acre has, in most instances, corrected the insufficiency. Eroded soils, high-lime soils with a pH of 7–8, peat and muck soils, and leached, sandy soils generally show a paucity of available zinc.

Most of the zinc in a soil is present in combined form either in organic complexes or in various minerals, and it is therefore not

[13] Powers, W. L., and Pang, T. S., *Soil Sci.*, **64**, 29–36 (1945).
[14] McMurtrey, J. E., Jr., and Robinson, W. O., *USDA Yearbook*, 807–829 (1938).

9.2 White bud or zinc deficiency in corn or maize. (*Courtesy of Fla. Agric. Expt. Station*)

readily available to plants, since they can take up only water-soluble or exchangeable forms of zinc. Soils have been found to vary widely in their capacity to supply the element, regardless of the total amount they contain. Parent rock material influences somewhat the total and available amounts of zinc in a soil, for example, limestone soils contain more zinc than do those derived from gneiss or quartzite. Bokde[15] published data on the zinc content of soils in India sampled from many

[15] Bokde, S., *Fertiliser News India*, **8** (3), 27–34 (1960).

9.3 Zinc deficiency in peaches. (*Courtesy of Fla. Agric. Expt. Station*)

parts of that country: available Zn varied from 0.50–6.05 ppm, with an average range of 1.79–4.57 ppm. The total content ranged from 20–95 ppm, with an average of 52.4–80.6 ppm.

Solubility Value

The solubility value established for certain compounds of zinc help one to understand some of the major factors affecting the availability of soil zinc at different pH values. For example, the solubility values

9.4 Zinc deficient peach leaves, showing the range from normal (left) to acute (right) (*Courtesy of Fla. Agric. Expt. Station*)

for zinc carbonate or hydroxide suggest that a soil having a high pH would usually contain a small amount of available zinc. In the case of soils characterized by a high content of hydroxyl $(OH)^-$ ions, it is difficult to get a crop response to an application of zinc due to the formation of zinc hydroxide. It is known that zinc nitrate is very soluble in water. This perhaps explains why there is a relationship between the amount of the nitrogen supply in the soil and the zinc supply; the presence of nitrates in the soil would significantly favor the formation of zinc nitrate, which may move with the drainage water and be leached out of the root zone or otherwise provide a supply of available zinc. Similarly, the presence of chloride or sulfate ions in the soil solution would favor the promotion of soluble zinc chloride or sulfate and the consequent increase in the amount of available soil zinc. Hence, when zinc is added to soils having a relatively low pH or a relatively high concentration of nitrate, chloride, or sulfate anions, the supply of soluble, available zinc would increase for many crops. The unfavorable effect of adding nitrate-chloride- or sulfate-containing fertilizers to soil could bring about a depletion of the native supply of available or exchangeable zinc through leaching. As previously pointed out, zinc ions remain quite immobile and, furthermore, most plants do not readily absorb the weak zinc ions which would explain why tree crops particularly respond better to foliar than to soil applications.

Slowly soluble zinc fertilizer materials must be at least as fine as 200-mesh for satisfactory performance, according to experiments by U.S. Department of Agriculture scientists[16]; they used two widely different soils under greenhouse conditions and Ranger alfalfa as the test crop. Low applications of frits (glass), although having less effect on the zinc content of the crop, increased the yield more than did a fine, crystalline zinc sulfate. Zinc ammonium phosphate, pulverant, supplied an adequate amount of zinc. Zinc chelate (Zn EDTA) increased the zinc content twice as much as zinc sulfate in neutral soil and up to six times as much in a calcareous soil. The influence of the chelated zinc on the crop zinc remained relatively stable over a 9-month period.

Liming Effects on Zinc Reactions in Soil

Adsorption of nutrients under different soil conditions was investigated in Finland. In one experiment soils were limed at 0, 8, 16, 24, 32, and 64 tons/hectare with limestone ($CaCO_3$) 7–9 years prior to the adsorption studies[17]. The soil cation exchange capacity, pH, and percent saturation were determined. Carrier free ^{65}Zn, ^{89}Sr, and ^{60}Co were the isotope cations for determining the amount of adsorption. Soil pH values greater than 7 increased the amounts of Mn, Cu, Ni, Co, and Mo extracted with NH_4OAc. The adsorption of ^{65}Zn and ^{60}Co without carrier increased up to a pH value of 6.5 and then remained constant. The adsorption of zinc decreased as the Zn concentration increased on unlimed soil, whereas on heavily limed soils zinc adsorption increased to a constant level. Robert S. Whitney* commented as follows on liming:

> "Probably no other practice affects soils as a plant growth medium as much as liming an acid soil. Every aspect of the plant's growth and life is touched in some way when acid soils are neutralized by liming. . . Of particular significance in this regard is the interaction and availability of plant nutrients."

[16] Lakanen, Esto, *TVA abstract*, No. 309 (February, 1968).
[17] Price, N. O., and Moschler, W.W., *J. Agr. Food Chem.*, **13** (2) (1965).
* Past President, American Society of Agronomy.

Residual lime in the soil caused significant changes in the composition of peanut foliage, soybean foliage, and orchard grass 7 and 9 years after the application of lime, according to an investigation at the Virginia Agricultural Experiment Station (AES)[17]. Copper, zinc, iron, cobalt, and manganese were significantly lowered in each plant, whereas molybdenum was increased.

The effect of liming on the availability of zinc and copper was studied at the University of California (Davis) under greenhouse conditions[18]. The soils were Yolo fine sandy loam and Yolo clay. The rates of lime were up to 50% $CaCO_3$. The concentrations of zinc and copper in the plant tissue were not markedly affected by increasing the rates of lime, particularly when the pH was not altered. A significant depression occurred in the plant growth on the Yolo clay as the lime rates increased.

Wear[19] reported a reduction in the concentration of zinc in sorghum tissues as a result of liming, although rates up to 4 tons/acre did not induce zinc deficiency symptoms.

Recent Reports from State Agriculture Experiment Stations

A brief summary of recent investigations on zinc deficiency reported by several state agricultural experiment stations (AES) will show the trends in research on this subject in the United States.

Numerous zinc deficiency problems have cropped up in the western half of the country, but the major areas reporting serious deficiencies seem to be located east of the Rocky Mountains. The need for zinc has been established in at least 31 states, according to the latest survey[20]. The crops most affected are corn (maize), field beans, sorghum, and potatoes. Zinc deficiencies occur least often in the intensively cultivated areas, which have received heavy dressings of farm manures and on alluvial soils. Temperature is another factor inducing deficiencies; cool, wet periods early in the growing season favor them, but with warmer weather later the symptoms very often disappear.

Colorado State University AES reported on tests in which soil was incubated at 43°C for one week; the result was that twice as much

[18] Brown, A. L., and Jurinak, J. J., *Soil Sci.*, **98**, 170–173 (1964).
[19] Wear, J. I., *Soil Sci.*, **81**, 311–318 (1956).
[20] Berger, K. C., "Introductory Soils," New York, Macmillan Co., 1965.

Zn was made available as when the soil was incubated at lower temperature for up to 6 weeks[21]. Fertilizer sources of zinc were also studied; the results showed that chelated zinc fertilizers, applied in bands, were generally more effective than inorganic sources. Chelates are compounds in which a metal ion combines with two or more sites of a single organic molecule and, thus bound, loses its ability to participate freely in some chemical reactions. Such metals are, accordingly, prevented from soil fixation, oxidation, and/or precipitation involvements. Synthetic chelates, although in themselves stable compounds, have the ability to release their bound metal ion to the extractive agents of the soil and plant. Otherwise such ion might become bound by soil agents into unavailable chemical compounds.

Tennessee Valley Authority Research Center using greenhouse pot experiments with local, zinc deficient soils sought the most effective method for incorporating various zinc compounds in the soil and their agronomic value. Corn (maize) was the test crop[22]. The results indicated that a close relationship exists between soil pH, the distribution in the soil of the zinc carrier, and its availability to the plant. Zinc deficiency was not serious below a pH value of 6, except on very sandy soils, and is intensified with a pH reading higher than 6 and also in calcareous soils. These tests also demonstrated that mixing the zinc carrier with the soil or applying it in granules or in bands of potentially acid fertilizers are effective practices. The amount of phosphorus in the corn plants was related inversely to the concentration of zinc. However, still unexplained remain the factors which govern this relationship.

Experiments conducted at the Alabama AES showed that lime, phosphorus, and potassium affect the amount of zinc absorbed by sorghum. Phosphorus application reduced the amount; without the applied P the plants contained 32 ppm; with an application of 175 lb P/acre, the plant content of Zn averaged only 19 ppm; and with 350 lb P/acre, the Zn concentration remained at about this same minimum value. Furthermore, without P application lime reduced the Zn uptake from 32 to 28 ppm but this was due rather to the higher pH value caused by the lime than to the lime specifically. These tests indicated that phosphorus had more influence on the absorption of Zn than the change in the soil pH; that potassium increased the Zn

[21] Lindsey, W. L., *Crops and Soils* (October, 1965).
[22] Terman, G. L., and Mortredt, J. J., *Crops and Soils* (October, 1965).

concentration in the plant but only when phosphorus was absent. Apparently potassium can affect Zn uptake by indirectly influencing the phosphorus-zinc relationship[23].

University of California (Davis) reported that its investigators found the following compounds of zinc of equal effectiveness in supplying nutrient zinc when they were thoroughly incorporated in the soil: zinc sulfate, zinc nitrate, zinc ammonium phosphate, zinc oxide, chelated zinc (Na_2 ZEDTA), and Rayplex Zn.

Some crops show no obvious leaf symptoms for zinc deficiency. This makes diagnosing, on the basis of plant growth, almost impossible. The amount of soil zinc extracted by ammonium acetate-dithizone solution may be useful in identifying areas likely to be deficient in Zn. Tentatively, the authors believe 0.5 ppm is a critical level below which a zinc response may be had with sensitive crop plants.

Greenhouse tests at Michigan State University AES[24] showed that zinc sulfate was the most effective source of nutrient Zn. Also, other results are the following: heavy application of phosphate fertilizer may induce a deficiency in available Zn; finely divided zinc oxide was as available as zinc sulfate, but granular zinc oxide was not satisfactory; generally, 3-4 lb Zn/acre in the inorganic form sufficed to correct Zn deficiency, but only about $\frac{1}{3}$ as much chelated Zn was of equivalent potency. (Uptake and yields from Zn EDTA, Zn HEEDTA, and Zn NTA were about the same on all plot locations). The tests also indicated that how the zinc compound is added in making the starter fertilizer is an important factor in its efficacy; adding ZnO or $ZnSO_4$ during the manufacturing process but prior to the granulation stage reduced their solubility in water, their availability to pea beans, and the final yield as compared to hand-mixing with the fertilizer at planting time; and that ZnO was more available to pea beans when it was incorporated into ammonium polyphosphate than into ammonium orthophosphate (pea beans are also called navy beans).

Another series of tests carried out at California University (Davis) developed the following results[25]: greenhouse experiments, using sweet corn as test plant, found that soil applied zinc remains fairly near where it is placed; much of the applied zinc remains in dithizone-extractable form, which correlates with availability to

[23] Wear, J. I., and Patterson, R. M., *Crops and Soils* (October, 1965).
[24] Ellis, B. A., *Crops and Soils* (October, 1965).
[25] Brown, A. L., Krantz, B. A., and Martin, P. E., *Soil Sci. Soc. Proc.*, **26**, 167–170, also **28**, 236–238 (1962).

plants; the mobility of applied zinc is greater in sandy soils; surface applications of zinc carriers frequently fail to correct a deficiency due to the immobility of the Zn.

Another greenhouse experiment[26] uses six different soils and sweet corn as the test plant; of the six soils, four had been proved Zn deficient by previous tests. Dithizone-extractable Zn had been determined on all soil samples. Plants in five of the six soils responded to applied zinc. Ten successive crops were grown on these soils. The rates of application were 0, 4, and 20 mg Zn/1,600 gm soil. Dry weight showed that the 4 mg Zn rate was adequate for six or seven successive crops. A portion of the applied Zn reacted with the soil to become unavailable, according to the dithizone extraction technique. All soils receiving 20 mg Zn and four to which 4 mg were added still had sufficient extractable Zn, after 10 crops had been grown on them, to supply the requirements.

Alfalfa was also grown in the same pots. Soil zinc increased the zinc concentration of the alfalfa, even on those soils considered to be markedly deficient for the Zn sensitive crops, since the concentration was invariably 20 ppm or higher (alfalfa is not in the Zn sensitive category).

Washington State AES investigation with potatoes by USDA scientists showed that zinc deficiencies that caused "fern leaf" symptom in Russet Burbank potatoes were invariably associated with applications of phosphatic fertilizers. There are several deficiency symptoms; young leaves cup upward and roll so much that the terminal growth looks like the unfolding fronds of certain ferns; stunted growth is another symptom. Potato plants exhibited deficiency symptoms by the end of the bloom period when grown on plots fertilized with 43% triple superphosphate at the rate of 100 lb/acre. Zinc was applied as the sulfate salt. It is not definitely known whether P interferes with zinc availability by preventing the entry of Zn into the plant or by blocking the movement of Zn from the roots to the leaves.

Studies in zinc deficiency at the Ohio Agricultural Research and Development Center were reported by Jones[27], who indicated that under Ohio conditions the need for application of micronutrients, particularly boron, copper, and zinc, to the corn (maize) crop has not

[26] Brown, A. L., Krantz, B. A., and Martin, P. E., *Soil Sci. Soc. Proc.*, **26**, 167–170, also **28**, 236–238 (1962).
[27] Jones, J., and Benton, Jr., *Ohio Agric. Res. and Devel. Center, Circular*, 149 (November, 1966).

been established. A 1964 summary of corn plant analyses showed that 5% of corn samples analyzed may have been deficient in zinc but none in boron or copper.

Furthermore, the effects of zinc treatments on leaf composition varied with sources and rates of application. Zinc carbonate $(ZnCO_3)$, at 5 lb Zn/acre, increased the zinc content of corn leaves when the plants were three to four feet tall, with the largest increase occurring during the early growth stage. Zinc sulfate $(ZnSO_4 \cdot H_2O)$ was as effective as the carbonate, as measured by the Zn concentration of the leaves, and in a comparison involving several zinc sources it was the most effective source in increasing the Zn content of the whole plant. NuZn and Sequestrene Zn 450 were less effective in this respect than zinc sulfate but more effective than zinc oxide.

Zinc in Grassland Herbage (Temperate Regions)

Dairymen and livestock men are greatly interested in techniques that will increase yields of dry matter (DM) from both cut and grazed herbage. To achieve maximal production of herbage, however, is not the only goal; its nutritional quality is of vital importance and must also be considered, since the ultimate purpose is the production of milk, meat, and wool. According to the authorities, animals require 14 mineral elements in their diets in addition to some 30 other components. For example, a dairy cow weighing about 1,100 lb and yielding about 3 gal of milk should have in its daily ration 50 ppm Zn in the dry matter of its diet[28], together with small amounts of the other known essential nutrients.

Grasses and legumes have about the same content of zinc, on the average about 15–60 ppm, and plants deficient in this element rarely contain less than 10 ppm[29].

The Zn requirement of cattle has been reported variously at 30–40 ppm, when the ration contains 0.3% Ca, with an increase of 16 ppm Zn for each additional 0.1% Ca. Zinc is generally considered nontoxic to animals[30].

[28] Agric. Research Council, London (1966).
[29] Hodgson, J. F., et al., *J. Agr. Food Chem.*, **10**, 171–174 (1962).
[30] Underwood, E. J., "Trace elements in human and animal nutrition," 2nd Ed., New York, Academic Press Inc., 1962.

Fertilizers can affect the zinc concentration of herbage. The Zn content of Coastal Bermudagrass was increased when applications of nitrogen fertilizer exceeded 400 lb/acre. Phosphate can reduce the availability of soil Zn to pasture and other crops. Liming may reduce the Zn content of herbage. Reith and Mitchell report[31] that applying 3 tons limestone to a soil of pH 5.9 reduced the Zn concentration in mixed herbage from 32 to 28 ppm.

A recent test in Australia involved annual pasture and crop plants from 21 species grown on a lateritic, gravelly sand. Three rates of a trace element mixture containing Cu, Zn, Mn, Mo, Co, Se, and I were applied. The plants were sampled at three stages of growth and at maturity. The Zn content decreased with the age of the plants, although it continued to rise throughout the growing season in all species except lupins, which showed significant declines after flowering. The Zn content varied widely among the species, with legumes and herbs having a higher concentration than cereals and grasses. It was apparent that species differ in a characteristic way in their power to absorb zinc, and this may help to explain why tests reveal differing degrees of susceptibility to Zn deficiency in the field[32].

Zinc Deficiency in Sugar Cane

Some Florida sugar cane fields have reported a zinc deficiency where an application of 30 lb $ZnSO_4 \cdot H_2O$/acre corrected the condition. Evans[33] described the symptoms of zinc deficiency in sugar cane:

"The most characteristic early symptom is a pronounced palish green color along the major veins. In marked contrast to iron and manganese deficiencies in which the striped chlorosis is interveinal, in zinc deficiency the striped effect is due to a loss of chlorophyll along the veins, although the interveinal area also becomes progressively paler with increasing deficiency. The distinct white-line appearance of the major veins appears due to the reduced chlorophyll in the bundle sheath of the main bundles. When the symptoms become acute there is evidence of veinal necrosis, and ultimately the growing point dies. Young tillers emerging are completely chlorotic in severe deficiency, and leaves soon become necrotic from the top downwards."

[31] Reith, J. W. S., and Mitchell, R. L., *Am. Soc. Hort. Sci. Fert. Problems*, **IV**, 241–254 (1964).
[32] Gladstone, J. S., and Loneragan, J. F., *Australian J. Agr. Res.*, **18**(3), 427–446 (1967).
[33] Evans, H. E., *Proc. 10th Cong. Intern. Soc. Sugarcane Tech.* (1959).

In the Natal (S. Africa) Cane Belt, du Toit[34] reports that Zn deficient cane becomes extremely stunted in growth, stooling is poor and lacks a healthy color, taking on a yellowish appearance with a tendency to striping. The soils in such cases were quite acid and low in phosphorus. In some experiments conducted at the Natal AES, zinc sulfate applications on problem soils gave dramatic results; the effect was seen in the increase in the growth and in the number of tillers. 50 lb of $ZnSO_4 \cdot H_2O$ was the dose per acre, and it was worked into the soil when the plants were about 8 months old. The normal color was restored and growth was stimulated.

The increases in the concentration of zinc in the plants four months after the application were as follows on a dry basis:

	Control ppm	Zn treated ppm
3rd leaf laminae	5.5	10.0
Top centimeter from growing point	60.0	130.0
Top 3 nodes and internodes	6.5	12.5

Heavy liming depressed the growth of the cane.

Zinc Deficiency in Tobacco

The average Zn content in various commercial varieties of tobacco is in the range of 51–84 ppm, according to Ward[35]. In cultural solution tests, tobacco plants receiving 0.0005 ppm or less Zn, displayed visual deficiency symptoms; leaves died, tissues matured earlier, and small tumors developed behind the growing root tips. Also determined was the cessation of meristematic activity in the root tips and in the cambium.

When Zn deficient subsoil was used in some California tests, the plants developed "little leaf" or rosetting, and this effect was prevented by the addition of zinc to the soil[36].

In a nutrient solution experiment[37] Zn deficient plants suffered

---

[34] du Toit, J. L., *S. African Sugar J.* (May, 1962).
[35] Ward, G. M., *The Lighter*, **11**(1), 16–21 (1941).
[36] Hoagland, D. R., et al., *Proc. Am. Soc. Hort. Sci.*, **33**, 131–141 (1936).
[37] Toui, Cheng, *Am. J. Botany*, **35**, 309–311, 172–179 (1948).

a decrease in auxin before any visual symptoms developed. When zinc was added, the free auxin and the enzyme-digestible, bound auxin increased within 2 days. Zinc is needed by plants for the synthesis of amino acid tryptophan and indirectly for the synthesis of auxin. It also functions to promote the formation of carotene and citrin in plant metabolism.

Major Zn Deficiency Symptoms in Plants (see Figures 9.1, 9.2, 9.3, 9.4).

Unlike phosphorus, zinc is not translocated from old to new tissues, and consequently, it exhibits various combinations of chlorosis, dieback, and rosetting in young growth. The evidence from numerous records indicates that visual zinc deficiencies occur mostly in fruit and nut plants and in corn (maize) as the chief field crop. The most common symptom is a yellowing or chlorosis of the leaves.

Deficiency in corn and sorghum shows up in the early stages of the growth cycle when the plant is only a few inches high, appearing first as yellow streaks in the leaves with a white to yellowish top. White spots may appear in the leaves or along the edges, and a portion of the marginal area may die; the entire plant may be stunted due to a shortening of the internodes.

In soybeans much of the same general symptoms occur as in corn, and in severe cases the plants become stunted, the leaves take on a yellow or pale green appearance, and all the lower, older leaves become brown and drop off. The plant may die prematurely.

The symptoms in citrus are as follows: the leaves show irregular, yellowish areas between the veins, and as new foliage develops, it becomes progressively smaller (the so-called "little leaf" disease).

In peaches the leaf symptoms are quite similar to those of citrus. In severe cases rosetting occurs, leaves die and drop off, young twigs die back for several inches, and the fruit is misshapen and reduced in quantity and quality.

In pecans the tree takes on a brownish or bronze appearance when seen from a distance, the nuts are small and poorly filled, and the yield is reduced. In severe cases, the leaves, twigs, and often larger limbs in the crown of the tree may die. Rosetting frequently is exhibited.

In tung trees bronzing of the foliage is characteristic; trees may become stunted and die within 2 or 3 years. The trees may lose their ability to resist the effects of low temperatures.

In cotton the interveinal areas of the leaves become chlorotic at first and then later the leaves become bronze and brittle, their margins turn up, and most of the affected leaves die.

Plant analysis is a helpful tool in determining Zn deficiency in tree fruits, nuts, and grapes and is particularly helpful in studying the relative effectiveness of different zinc treatments of all crops. The critical concentration in the plant varies with the species, the stage of growth, and the tissues. A general guide for crops like corn (maize), tomatoes, and beans is to consider a concentration of 12–18 ppm dry weight as the critical range. Soil analysis may be helpful in identifying soil areas likely to be deficient except in the case of grapes, tree fruits, and nuts. On the basis of greenhouse and laboratory studies (in California), a tentative critical level of 0.5 ppm Zn is suggested, below which a growth response to applied zinc might be expected on sensitive crops. Table 9.3 summarizes recommendations by states of zinc control applications on specified crop plants.

Zinc in Animals

Zinc plays an essential role in the nutrition of animals and man. A complete absence of this element in their metabolism can be fatal. Although the amount required may be only minute, it must be available for normal growth. Animal feedstuffs ordinarily contain an adequate amount. In cases of proved deficiency the usual procedure is to add zinc sulfate to the feed mixture. A typical level of Zn in feeds is about 50 ppm (0.005%) by weight, equivalent to 50 gm Zn/metric ton (1.6 oz/short ton).

Deficiency Symptoms

A moderate deficiency in the ration may show up as follows: retarded growth, bone and joint disorders, skin diseases, disorders in feathers and hair, delayed sexual maturity, sterility, and even death.

In recent years a disease in swine known to veterinarians by the name "parakeratosis" was traced to Zn deficiency in the metabolism.

TABLE 9.3   ZINC RECOMMENDATIONS BY STATES

STATE	CROPS	RECOMMENDATIONS FOR APPLICATION OF ZINC
Alabama	pecan	10–15 lb $ZnSO_4$ on 20 year old tree exhibiting rosette. 4–5 lb $ZnSO_4$/tree/year maintenance application.
	peach	No general recommendations.
	corn	10 lb $ZnSO_4$/acre.
Arkansas	pecans	No general recommendations.
Arizona	pecans	No general recommendations.
California	All tree crops and field crops	Tree crops: 10–25 lb $ZnSO_4$/ 100 gal as dormant spray. Foliar spray: 5 lb ZnO/100 gal Field crops: 10–20 lb Zn/acre from $ZnSO_4 \cdot H_2O$.
Colorado*	corn	5–10 lb Zn from $ZnSO_4 \cdot H_2O$/ acre of 0.5–1 lb Zn from EDTA Zn.
	beans sorghum fruit trees	
Florida	Tree crops: citrus, tung, pecans, peach, guava, avocado, mango	3 lb $ZnSO_4$, plus 1 lb hydrated lime/100 gal water applied as dormant or post-bloom spray.
	Vegetable and field crops: corn, cowpeas, sorghum, soybeans, celery, beans, peanuts, tomatoes, squash, ramie, potatoes, clover and pasture.	Prevent: .2–.516 lb ZnO (as $ZnSO_4$) in fertilizer every 4–5 years. Correcting: 2 lb $ZnSO_4$/ 100 gal H_2O or zineb at 2 lb/ 100 gal.
	Ornamentals: orange-jasmine, wax privet, loquat, silk oak, Hetai. catalpa, American elm.	Prevent: .2–.590 lb ZnO (as $ZnSO_4$) in fertilizer every 4–5 years. Correcting: 2 lb $ZnSO_4$/ 100 gal H_2O or zineb at 2 lb/ 100 gal.

TABLE 9.3 (continued)

STATE	CROPS	RECOMMENDATIONS FOR APPLICATION OF ZINC
Georgia	pecans	Zinc sulfate, $\frac{1}{2}$ lb/in. trunk diameter.
	peaches	Zinc sulfate, 2–3 lb/tree after symptoms have been observed.
	corn	$ZnSO_4$, 10–20 lb/acre.
Indiana	corn	2 lb Zn (applied as plow down). 1 lb (applied in row). Carrier: $ZnSO_4 \cdot H_2O$.
Iowa	corn soybeans	10–20 lb Zn/acre. Carrier: $ZnSO_4 \cdot H_2O$.
Kansas	corn sorghum	10–15 lb Zn/acre. Carriers: $ZnSO_4 \cdot H_2O$, Zinc chelate.
Kentucky	corn	10 lb $ZnSO_4$ in row at planting. 20 lb/acre if pH and phosphate in soil are high.
Louisiana	citrus	3 lb $ZnSO_4$ and 1 pound of hydrated lime/100 gal water (foliar spray). Chelated zinc to soil.
	pecan	3 lb $ZnSO_4$ and 1 pound hydrated lime/100 gal water (foliar spray). Chelated zinc to soil.
	tung	3 lb $ZnSO_4$ and 1 pound hydrated lime/100 gal water (foliar spray). Chelated zinc to soil.
Michigan	beans	(1) Alkaline, organic soils: 2–4 lb Zn from $ZnSO_4 \cdot H_2O$. (2) 3–4 lb Zn on poorly drained soils, banded with fertilizer. Carrier: $ZnSO_4 \cdot H_2O$, ZnO.

TABLE 9.3 (continued)

STATE	CROPS	RECOMMENDATIONS FOR APPLICATION OF ZINC
Maryland	corn	10 lb Zn/acre from $ZnSO_4 \cdot H_2O$.
Minnesota*	corn	Excessive free lime in soil. 10 lb. Zn/acre from $ZnSO_4 \cdot H_2O$ or 0.5 lb Zn from Na_2An (chelate).
Mississippi	pecans tung corn	5–10 lb $ZnSO_4$/tree. 5–10 lb ZnSO/tree. 10 lb $ZnSO_4$/acre.
Montana*	cherries apples	15 lb/acre Zn from $ZnSO_4 \cdot H_2O$.
Nebraska*	corn sorghum soybeans castor beans fruit sugar beets	5–20 lb Zn from $ZnSO_4 \cdot H_2O$.
N. Carolina	peaches	6 oz $ZnSO_4$/100 gal water applied as spray every 3 weeks from May 1 to July 15.
N. Dakota*	corn potatoes	15 lb Zn/acre from $ZnSO_4 \cdot H_2O$.
Oklahoma*	pecans	4 lb Zn/acre.
Oregon*	corn lima beans all tree fruit cherries pear barley	5.5–11 lb Zn from $ZnSO_4$. 5.5–11 lb Zn from $ZnSO_4$. 2 lb/100 gal from chelate. 12–15 lb/100 gal from $ZnSO_4$. 5 lb/100 gal from $ZnSO_4$. 15 lb Zn/acre.

TABLE 9.3 (continued)

STATE	CROPS	RECOMMENDATIONS FOR APPLICATION OF ZINC
S. Carolina	corn	10 lb $ZnSO_4$/acre.
	peaches	Neutral zinc according to manufacturers' recommendation for bacterial spot—secondary effect as fertilizer.
	pecans	10–12 lb $ZnSO_4$/mature tree. 2–5 lb annually to prevent rosette.
	vegetables	5 lb $ZnSO_4$/acre.
Tennessee	pecans	General recommendations.
	corn	10–20 lb $ZnSO_4$/acre.
	sorghum	No general recommendations.
	sudan grass	No general recommendations.
Texas	citrus	Foliar sprays containing zinc.
	peaches	Neutral zinc for bacterial spot control.
	pecans	Foliar spray of $ZnSO_4$—3 lb/100 gal water with casebearer spray when leaves are $\frac{1}{3}$ grown.
Utah*	fruit	10 lb Zn from $ZnSO_4 \cdot H_2O$/acre.
Virginia	peaches	No general recommendations.

* Deficient levels less than 16–20 ppm Zn in plant, oven-dry basis.
Source: *Plant Food Rev.* (1963).
    Soil Sci. Soc. Am. (1965).

Supplemental zinc in the ration has eliminated the condition. Zinc sulfate ($ZnSO_4 \cdot H_2O$) is an approved source of the element and a feed containing 0.6 lb/ton of feed is recommended by authorities.

A forage having a zinc content of only 18–42 ppm dry weight may cause a deficiency in cattle grazing it. When Zn deficiency was induced in tests with Holstein calves, the symptoms were inflammation of nose and mouth tissues, loss of weight, slight swelling of hind feet, and hard, dry skin on the body and head.

The common symptoms in chickens and turkey include brittle feathers and bones, difficulty in walking, retarded growth, scaly skin, delayed maturity, poor hatchability of eggs, lowered egg laying. The supplemental diet to correct the deficiency is as follows: chicks on a casein diet require 12–14 ppm Zn or 0.07–0.08 lb $ZnSO_4 \cdot H_2O$ per short ton of feed. If on a soybean diet add 27–29 ppm or 0.16–0.17 lb of the zinc sulfate per short ton of feed.

Young chicks should be able to tolerate up to 1,000 ppm Zn as the sulfate form or up to 6 lb/short ton. The recommended level is 50 ppm or 0.3 lb $ZnSO_4 \cdot H_2O$/short ton of feed.

Turkeys require more Zn in the diet than chickens. A recommended level is 100 ppm Zn as a supplement obtained by adding 0.6 lb $ZnSO_4 \cdot H_2O$/short ton of feed.

Conversion Chart

Amounts of ZnO and $ZnSO_4 \cdot H_2O$ to give specified amounts of elemental zinc (Zn) are as follows*:

Zinc	Zinc Oxide ZnO	Zinc sulfate $ZnSO_4 \cdot H_2O$
1	1	3
5	6	14
10	12	28
20	28	55
25	31	69
30	37	82
40	50	110
50	62	137
60	75	165
70	87	192
75	93	206
80	100	220
90	112	247
100	124	275

* These figures apply to whatever units are employed. For example, to obtain 10 lb of Zn, it is necessary to employ 12 lb of ZnO or 28 lb of $ZnSO_4 \cdot H_2O$. Instead of pounds it could be grams or parts per million; 10 gm Zn needed would be 12 gm of the oxide or 28 gm of the sulfate.
Source: Sherwin-Williams Co.

10

MOLYBDENUM

Although molybdenum (Mo) has been a comparatively recent addition to the group of nutrients essential to plant life, it is now well established in that role. Mo is involved in soil-plant-animal relationships. It is concerned with Mo deficiency in plants and with Mo toxicity in animals. Heaviest among all the metallic elements required by plants, it is also the one element required in least amount. In 1932 Ter Meulen[1], an early investigator of the Mo content of soils and plants, reported that Mo was a common constituent of soils and plants. He found that fertile soils contained, on average, from 0.1–0.3 ppm Mo, and barren sands, 0.005 ppm, while the content of plant tissues varied from trace to 9.0 mgm/kg dry matter, with legumes containing relatively higher amounts than nonleguminous plants, and with mere traces being found in the tissues of fruits and vegetables, and in the leaves and wood of trees. In a later report Beeson[2] stated that the concentration of Mo in plant tissues may vary from less than 0.1 ppm to more than 300 ppm Mo without any apparent effect on plant growth. However, if plants containing more than 15–20 ppm Mo are eaten by

[1] Meulen, H. ter, *Nature*, **130**, 966 (1932).
[2] Beeson, K. C., *Soil Sci. Soc. Am. Proc.*, **25**, 227–232 (1961).

livestock, the animal develops a conditioned copper deficiency called molybdenosis.

World interest in molybdenum really began in 1942, when Mo assumed significant importance in agriculture. In that year A. J. Anderson, an Australian scientist, demonstrated Mo deficiency in some low producing, hill land, clover pastures near Adelaide, South Australia by a series of field experiments. By applying as little as $\frac{1}{16}$ oz Mo to the acre, the deficiency was corrected[3]. Furthermore, he showed that the principal effect of the applied Mo was to reinstitute normal performance by the nitrogen-fixing, root nodular bacteria on the clover roots. However, as early as 1930 Bortels in Germany found that Mo was associated with the fixation of atmospheric nitrogen by *Azotobacter chrococcum*, and in 1937 he reported the first experiments in which a legume responded to applied Mo fertilization. He also noted that alfalfa plants treated with Mo produced a significant increase in seed[4]. Anderson's research further dramatized the inter-dependence between the symbiotic nodular organisms and legumes, and the association of Mo with the activities of these nitrogen-fixing bacteria. A large commercial market for the element was to develop, making a significant impact on agriculture.

The Australian discovery stimulated interest in European and American scientific circles. Reports of deficiencies in other crops besides legumes followed in quick succession. It was evident that plant diseases amenable to correction by applications of Mo had been known for many years to plant pathologists but not the true nature of the cause. For instance, the "whiptail" disease of cauliflower had been described as early as 1924 as being due to some nutritional disturbance correctable by liming the soil. We know now that molybdenum may in many cases be converted by soil agents into unavailable forms, particularly in acid soils, but that liming to a higher pH value will release adequate amounts of Mo to satisfy the plant requirement Workers in Australia and New Zealand reported spectacular increases in crop yields, some as high as tenfold, brought about by Mo fertilization supplementing superphosphate applications.

Until 1939 the status of Mo as a plant nutrient was generally accepted, but the true scientific bases of its essentiality to plants and animals was firmly established in that year by Arnon and Stout,

[3] Anderson, A. J., *J. Australian Inst. Agr. Sci.*, **8**, 73–75 (1942).
[4] Bortels, H., *Arch. Mikrobiol.*, **1**, 333–342 (1930).

10.1 Molybdenum deficiency in cauliflower. "Whiptail" effect. (*Courtesy of Climax Molybdenum Co.*)

scientists at the University of California[5], and their results were confirmed by Piper in 1940[6]. Arnon and Stout showed that tomato plants could not complete their growth cycle, when growing in their highly purified water culture, unless Mo was added. An application of 0.01 ppm Mo permitted normal growth. Whiptail disease of cauliflower and broccoli was demonstrated in 1945 by New Zealand workers

[5] Arnon, D. I., and Stout, P. R., *Plant Physiol.*, **14**, 599–601 (1939).
[6] Piper, C. S., *J. Australian Inst. Agric. Sci.*, **6**, 162–164 (1940).

to be a result of Mo deficiency correctable by spraying the foliage with a solution of molybdenum salts or by Mo additions to the soil. Several other investigations between 1932 and 1945 convinced the scientific world that Mo is essential in the normal life processes of micro-organisms, higher plants, and animals.

Chemical Relationships

As already noted, molybdenum is the heaviest among all the chemical elements required in the nutrition of plants. Its atomic number is 42 and atomic weight 95.95. In the Periodic Table it is the third member of the subgroup of which chromium and tungsten are the others. All members are metals and they have a chemistry complicated by the existence of several oxidation states ranging from 2^+ to 6^+ and by the formation of many complex ions including some oxyions.

In nature molybdenum occurs as the mineral molybdenite (MoS_2). When heated in air MoS_2 is oxidized to the trioxide (MoO_3), in which Mo with a valence of 6 has its most important oxidation state. The trioxide is acidic and dissolves in basic solutions to form a complicated series of oxyanions called molybdates of which MoO_4^{2-} is the most common. More complicated molybdates are known and are called polymolybdates, some of which have up to 24 Mo atoms.

Other primary minerals which contain some molybdenum are wulfenite ($PbMoO_4$), powellite ($CaMoO_4$), and ferromolybdite ($FeMoO_4 \cdot 7\frac{1}{2}H_2O$). In acid soils Mo occurs adsorbed or included in crystals of hydrous iron and aluminum oxides. When adsorbed on the colloidal complex of the soil, it may be present as exchangeable ion able to react with other anions, such as PO_4, or be in solution as simple salts. Mo is believed to combine with iron in acid soils to form insoluble salts which release Mo very slowly for plant uptake, and this may be the reason why most soil deficiencies occur in strongly acid soils. Liming such soils raises the pH value upward to the neutral or alkaline levels and causes the release of enough Mo to supply the crop needs, assuming of course that the total amount of molybdenum is sufficient.

Mo in Soils

A systematic survey to determine the total Mo content of agricultural soils in the United States was conducted by Robinson and associates, and the results were published in 1951[7]. They reported that in about 95% of the samples examined, the content ranged from 0.6–3.5 ppm Mo. This range, they cited, compared with Russian and Argentine investigations in which the content varied from 2.0–2.6 ppm. In 18 of the U.S. samples the content ranged from 0.8–3.3 ppm. The published records show some soils in France have a content varying from 4.3–6.9 ppm Mo, and Hawaiian soils, 8.9–73.8 ppm. It must be pointed out that although these contents of total Mo in soils are interesting, they do not disclose the amount of Mo that is available to plants, which is the true measure of agricultural significance. Barshad[8] working with California soils having excessive amounts of Mo described four possible main forms: (1) water-soluble; (2) combined with organic matter; (3) exchangeable anion $MoO_4{}^{2-}$, adsorbed on colloidal complex; and (4) insoluble, held in the crystal lattices of minerals.

The pH of a soil is acknowledged by the experts as being the most influential factor in affecting the availability of Mo to plants. Mo deficiency occurs most often on acid soils. Soil clays adsorb Mo ions from acid solution, and such ions are in turn released to alkaline solutions[8]. Mo deficiency has been found on highly weathered limestone, on soils with a high content of iron and magnesium, and on coastal, sandy soils. Unlike the other micronutrients, the availability of molybdenum increases as the soil pH approaches neutrality or goes above it. Experience in Australia shows that an application of 2–3 oz of Na_2MoO_4/acre is about as effective in correcting Mo deficiency as an application of 2 tons/acre of limestone and more effective if the molybdate is supplemented with a dressing of about 225 lb/acre of limestone[9].

On some soils having a high content of Mo, it may follow that, by applying a high rate of lime, an excessive concentration of Mo could occur in legumes growing on such soils, due to the presence of the element in an available form resulting from the neutral or alkaline soil condition[10]. Livestock grazing on such legumes might develop molybdenosis.

Other factors than pH are known to cause a deficiency, among which may be mentioned the presence of significant amounts of other heavy metals in the soil solution such as copper, manganese, zinc, and

[7] Robinson, W. O., et al., *Soil Sci.*, **72**, 267–274 (1951).
[8] Barshad, I., *Soil Sci.*, **71**, 297–313 (1951).
[9] Cullen, N. A., *New Zealand Soil News*, 41–46 (1953).
[10] Massey, H. E., et al., *Soil Sci. Soc. Am. Proc.*, **25**, 161–162 (1961).

nickel; their presence seems to increase the plant's need for molybdenum. Salts containing the sulfate ion (SO_4^{2-}) may decrease the amount of Mo a plant can take up from an acid soil, whereas PO_4^{3-} ions seem to aid in the uptake of Mo ions.

The Mo ion is immobile in the soil, being in this respect very similar to the PO_4^{3-} ion. If it is not removed by cropping or leaching or if not converted into insoluble forms in the soil, it could accumulate to the extent of becoming toxic to plants and to animals feeding on such plants.

Effects on animals

Mo in Plants

Molybdenum is an essential plant nutrient. The amount required by plants varies with species. The concentration of Mo may vary from less than 0.1 ppm to more than 300 ppm. Plants vary in their ability to extract molybdenum from the soil solution, and this may explain the

10.2 Legumes require molybdenum for fixation of nitrogen. The pale white clover plants at the right, which are deficient in molybdenum, contrast with the healthy plants at the left, grown with sufficient molybdenum. (*Courtesy of Climax Molybdenum Co.*)

variation in the concentration. The average amount of Mo found in agricultural soils generally is reported to vary from 0.5–3.5 ppm, but where the content of available Mo is high, some species will absorb more proportionately and reach the 300 ppm content[11]. Roots contain a greater proportion of the element than aboveground parts or seeds. For example, the roots of red clover on high Mo containing soils will contain on average 53 ppm, while the leaves have 29 and stems 28 ppm. In another experiment Mo deficient red clover leaves had only about 0.1 ppm, while alfalfa grown under the same conditions had 35 ppm in its roots, 7 ppm in its leaves, and 4.3 in its stems[11]. Pea plants contain an average of about 0.4 ppm Mo when deficient, but if they have over 0.7 ppm they may be considered adequately supplied[11]. Nodules from alfalfa roots were found to contain 5 to 15 times as much Mo as was present in other root tissues.

Anderson found that clover in a grass–legume sod responded to applied molybdenum when its content of the element was 0.3 ppm or less[12].

Lemon leaves deficient in Mo were found to contain 0.01 ppm, whereas normal leaves contained 0.024 ppm. The clover apparently required about 30 times more Mo than the lemon leaves. Legumes require relatively larger amounts of Mo than nonlegumes for normal growth. The nitrogen-fixing bacteria in the root nodules and present in all legumes utilize Mo in the process of fixing gaseous nitrogen, and this explains why legumes have a relatively greater requirement for Mo.

Function in Plants

As previously mentioned, Bortels showed the close association of Mo with the symbiotic root nodular bacteria and assumed it acted as a catalyst. Early research demonstrated that the fungus *Aspergillus niger* required more Mo when it was supplied nitrogen in the nitrate form. This indicated that Mo was possibly involved in the reduction of nitrates within the plant tissues[13]. Plants absorb nitrogen in the nitrate or ammonia form. Before it can be utilized in the plant metabolism, the NO_3^- nitrogen must be reduced to the ammonia form, and this is achieved by enzymes.

[11] Berger, K. C., "Introductory Soils," New York, Macmillan Co., 1965.
[12] Anderson, A. J., *J. Australian Inst. Agr. Sci.*, **8**, 73–75 (1942).
[13] Steinberg, R. A., *J. Agric. Res.*, **52**, 429–448 (1936).

10.3   Lettuce grown in nutrient solutions shows effect of molybdenum on growth. Stunted plant at the right was given all necessary elements except molybdenum. The healthy plant at the left received the same elements and molybdenum. (*Courtesy of Climax Molybdenum Co.*)

Research has definitely shown that the enzyme nitrate reductase catalyzes the reduction of the nitrate and that Mo is the prosthetic or indispensable group of the enzyme; its place cannot be occupied by any other element. Since Mo is involved in the fixation process of the root nodular microorganisms and in the nitrate reductase enzyme which controls the reduction of inorganic nitrate to a form the plant can utilize in building protein, Mo has been designated the key element in plant nitrogen metabolism.

Molybdenum has other functions in the plant; it is required in the synthesis of ascorbic acid; in making iron physiologically available within the plant; and in alleviating plant injury caused by the presence of excessive amounts of the elements copper, boron, nickel, cobalt, manganese, and zinc[14].

[14] Millikan, C. R., *Australian Inst. Agr. Sci.*, **13**, 180–186 (1947).

Mo Deficiency and Symptoms

Investigators had observed in the early research on the physiology of molybdenum that plants deficient in or lacking the element appeared to accumulate nitrate or oxidizing compounds in their tissues. This situation was in time explained by experimental evidence; plants and bacteria require and utilize Mo to reduce nitrate to the ammonia form by means of the nitrate reductase enzyme. This is a function of molybdenum that is well established and accepted by the plant physiologists. All Mo deficient plants grown with nitrate have shown similar symptoms[15], including chlorosis (yellow-green, yellow, or orange mottling of leaves), followed by marginal curling, wilting, and finally necrosis, and withering of the leaf. Symptoms usually appear in the older leaves first and then show up regularly in younger leaves until the growing point is killed, flowers withered or suppressed and eventually the plant dies. Rapid recovery is made possible by the timely application of Mo sprayed on the foliage or added to the soil.

The deficiency symptoms are associated with reduced yields and the accumulation of nitrate, which may increase to as much as 15% of the total dry material and with an abnormally low content of protein, total soluble nitrogen, and chlorophyll. In the brassica group of plants (cauliflower, broccoli), additional deficiency symptoms are displayed, such as the familiar "whiptail" disease. In this case the symptoms start always as one or more small rounded, translucent areas between major veins near the midrib of one or two young leaves and become yellow or totally chlorotic and then perforate, and as the leaf expands with growth, these perforations become enlarged, irregular holes. Maryland scientists have reported that the amount of molybdenum needed to produce the "whiptail" symptoms in plants grown in sand culture is between 0.000005 and 0.00005 ppm.

Mo deficiency in tomato and cauliflower causes a drop in the plant's content of ascorbic acid, and the effect of the decrease is proportional to the supply of Mo and independent of the supply of nitrogen. It also affects the content of total and reducing sugars[16].

At present the experts seem agreed that Mo is specifically needed only in the nitrate reductase enzyme system.

[15] Hewitt, E. J., *Soil Sci.*, **81**, 3 (1956).
[16] Hewitt, E. J., *J. Sci. of Food and Agric.*, **8**, 56–57 (1957).

10.4 Molybdenum deficiency of orange leaves, known as "yellow spot". (*Courtesy of Citrus Expt. Station, Florida*)

Blue-Chaff disease of oats is, according to Fricke[17], a result of Mo deficiency. Heavy liming or adding 1 ounce of Mo to the acre corrected the condition in Tasmania.

Hewitt and Jones[18] have described deficiency symptoms in 21 crop plants. Most of them seem to follow the same pattern; interveinal mottling and cupping of the older leaves followed by necrotic spots at leaf tips and margins. In alfalfa and other legumes the leaves become bright green, growth is retarded, and in time the older leaves become scorched and drop off prematurely. This condition resembled that

[17] Fricke, E. F., Tasmania Dept. Agric.
[18] Hewitt, E. J., and Jones, E. W., *Ann. Report Long Ashton Res. Sta.*, 81–90 (1948).

caused by an inadequate supply of nitrogen, which in a way it is, since under the circumstances the symbiotic nodular bacteria fail to grow and to fix atmospheric nitrogen.

"Yellow spot," a Mo deficiency disease of citrus, was known at least 40 years prior to the discovery in 1952 that it was caused by an insufficient supply of molybdenum.

Berger[19] surveyed deficiency of Mo returns in the United States and reported them by crops as follows: alfalfa in 13 states; cauliflower, 9; clover, 6; soybeans, 3.

The Soil Science Society survey[20] gives the following state reports, by crops, of Mo deficiencies:

North Carolina—cauliflower, lettuce, cantaloupe
South Carolina—alfalfa
Georgia—alfalfa, soybeans, crimson clover
Florida—citrus, cauliflower, sweet potato, broccoli, tomato, hibiscus
Louisiana—cauliflower, soybeans
Arkansas—soybeans
Mississippi—cauliflower, soybeans
Virginia—alfalfa
West Virginia—cauliflower
Michigan—onions, lettuce, cauliflower, spinach, red clover
Indiana—soybeans
Kansas—soybeans
Idaho—peas
Oregon—cole crops, forage crops
Washington—alfalfa, sweet clover, peas
California—alfalfa, peas.
(see Figures 10.1, 10.2, 10.3, 10.4).

Toxicity to Animals

Known for at least a century, an animal disease called "teart," associated with certain pasture areas in England, was finally traced to molybdenum toxicity. Teartness affects ruminants, especially cows when they are fresh and calves; it is characterized by extreme diarrhea, loss in weight, and reduced milk yields. The condition is now designated by the

[19] Berger, K. C., ibidem.
[20] Survey by Soil Test Comm. of Soil Sic. Soc. (March, 1965).

word, molybdenosis. In animal nutrition molybdenum attains importance primarily because of this disease.

In 1938 Ferguson and associates[21], studying a local disease of dairy cattle in Somerset, England, were able to link an excessive intake of molybdenum with the disease then known as teart. The forage in the affected area contained from 30–80 ppm Mo, while that from healthy areas contained only about 5 ppm, on a dry matter basis.

Forage having an excessive amount of Mo is found in the Florida Everglades and on the alkaline soils of some areas in the western part of the United States. Mo is widely distributed in soils. Plants require so very little of it in their normal metabolism. Availability of Mo to plants increases with pH from 6.5 to 7 and above. The levels at which Mo becomes toxic to ruminants will vary with the animal and the diet and particularly with the amount of copper and the SO_4^{2-} ion present in the diet and with the animal's physiology. The critical level of Mo for ruminants is generally considered to be between 5 and 10 ppm of Mo. It has been shown that Mo has a definite antagonistic effect on the metabolism of copper; animals grazing on herbage low in Mo content may develop copper toxicity if abnormal amounts of copper are present, or if the pasture content of Mo is relatively high, animals may develop a copper deficiency. Also, if the amount of sulfate ion taken in by the animal is increased beyond a normal level, it will cause a decrease in the amount of Mo that is retained by the animal.

A. T. Dick reviewed the literature on molybdenum nutrition (ruminant and nonruminant) and included a bibliography of 392 references to the subject[22]. He shows that when sheep grazed pastures of low Mo content (less than 0.1 ppm Mo, dry weight), copper rapidly accumulated in their livers, and this can lead to chronic copper poisoning and subsequent death of the animal. Conversely, under certain conditions when the diet contains about 5 ppm Mo, sheep may develop a clinical copper deficiency. The toxic effects of Mo in ruminants can be prevented by increasing the copper intake with a suitable copper compound, such as copper sulfate.

Mo in Grassland Herbage

The Mo contents in pasture species from various parts of the world have been reported as ranging from 0.01 to upward of 200 ppm

---

[21] Ferguson, W. S., et al.. *Nature.* **141**, 553 (1938).
[22] Dick, A. T., *Soil Sci.*, **81**, 299 (1957).

with the usual range, however, of about 0.1–4 ppm. Underwood[23] states that grasses usually have a higher Mo content than clovers, but Hemingway[24] found that in 2 out of 3 years white clover contained an appreciably higher content than grass from the same swards. He also found that levels in grass rose from 1.3–2.3 ppm between the first and fourth silage cuts on plots which received no nitrogen, whereas clover contents of Mo were much more uniform throughout the year.

Excessively high contents of Mo in herbage are present mainly on soils derived from certain shales and slates which have more than average contents of Mo. Alkaline soils and some organic soils are also liable to produce herbage with high Mo content.

Mo deficiencies occur very frequently in herbage grown on podzolic soils, calcareous sandy soils, and serpentine soils which have a low total content of Mo, on soils with high iron content, and on soils with a high anion exchange capacity combined with low pH values where availability is low[25].

Hemingway[24] studied the effects of NPK fertilizers on the Mo content of herbage. He reported that ammonium sulfate caused a decrease of about 59%, whereas superphosphate and potash (KCl) caused irregular and small effects. Ammonium sulfate undoubtedly influenced the pH toward acidity values, and the sulfate ion is known to depress the plant's uptake of Mo. Davies[25] reports that the phosphate ion increases the uptake of Mo.

Reith and Mitchell reported in 1964[26] that a single application of lime may increase the Mo of mixed herbage 2 to 3-fold and on some soils may raise it to a level above 5 ppm which may harm cattle feeding on it. These authors give the following average content of trace elements for agricultural liming materials: Mn, 100–1,000 ppm; Zn, 25 ppm; Cu, 1–10 ppm; Co, 5 ppm; B, 1–20 ppm; Mo, 1 ppm.

Intensive grassland production is bound to lower the soil's reserves of trace elements. In soils where the total amount of available supplies is low to begin with, deficiencies will necessarily be greater when high yields of herbage are cut and taken away rather than grazed in place. The reduction in availability of the metallic trace elements

[23] Underwood, E. J., "Trace elements in human and animal nutrition," 2nd Ed., New York, Academic Press, 1962.
[24] Hemingway, R. G., *J. Brit. Grassland Soc.*, **17**, 182–187 (1962).
[25] Davies, E. B., *Soil Sci.*, **81**, 209–221 (1956).
[26] Reith, J. W. S., and Mitchell, R. L., *Am. Soc. Hort. Sci.*, **IV**, 241–254 (1964).

and the increase in availability of molybdenum that can be caused by heavy liming is perhaps of more significance.

If it is assumed that the Mo content of cut herbage is 0.20 ppm, the amount of Mo removed in 5,000 lb dry matter will be 0.016 oz; if the content of Mo is assumed to be 3 ppm, the amount of Mo removed in 10,000 lb dry matter is 0.48 oz[27].

Corrective Measures

Molybdenum salts may be applied to effectively correct a deficiency either as a solid to the soil or by spraying an appropriate solution on the foliage or by treating the seed. The most commonly recommended compound used for this purpose is sodium molybdate, dihydrate, applied at a rate varying from $\frac{1}{16}$th oz to 4 lb/acre, depending on the established needs. The high rate (4 lb/acre) is used almost exclusively on the manganiferous soils of the Hawaiian Islands. Also used, but less frequently, are ammonium molybdate and molybdenum trioxide. These chemicals are applied mixed with fertilizer materials or with limestone. Climax Molybdenum, the manufacturer, recommends that Mo be incorporated into solid fertilizers in either of the following ways:

(1) Dissolve MoO or Na_2MoO_4 in the sulfuric acid to be used in acidulating phosphate rock to produce superphosphate.

(2) Mix the Mo compound mechanically with the solid NPK fertilizer. This must be very carefully done to assure uniform distribution of the Mo salt.

(3) Spray a solution of Na_2MoO_4 into the finished fertilizer as it moves along on the conveyor belt to storage or bagger.

The Australian practice for treating forage crops is to apply 2.5 oz/acre of the Mo salt. Florida citrus growers find the foliar spray method satisfactory; Mo is readily absorbed through young leaves and because the amount applied to the acre is small, the spray method of application of either Mo alone or with fertilizers and often pesticides has proved effective and economical.

Application on an acid soil to correct deficiency is best made by mixing the Mo salt with a lime material, or by adding sodium molybdate (39% Mo) or molybdic oxide (47.5% Mo) to a mixed fertilizer and

[27] Whitehead, D. C., "Nutrient minerals in grassland herbage," Comm. Agric. Bureau, 1966.

applying the mixture at a rate to supply the sodium form at from 2–8 oz/acre as required or at a slightly lower rate if the oxide is used.

For legume seed treatment use a commercial Mo material known to be compatible with the legume inoculant. About 1 oz of the material mixed with the seed is sufficient to plant from 1–4 acres. This seed treatment has worked effectively in the case of soybeans and other large seeded legumes.

In foliar spray applications either sodium molybdate, molybdic oxide, or ammonium molybdate (54% Mo) is made up in solution form and sprayed at a rate to supply 2–8 oz/acre of the salt. For the control of citrus leaf spot the usual rate is 1 oz of either of the salts dissolved in 100 gallons of water and applied at 10 gallons of the solution per tree.

Sodium molybdate may also be dissolved in the liquid fertilizers, 8-24-0 and 11-37-0, in amounts that are adequate to meet the need.

The official recommendation in New Zealand is to add Na_2MoO_4 at the rate of 1 lb/ton of fertilizer. It may be added in solution either during the mixing of ground phosphate rock and sulfuric acid or sprayed on the superphosphate. The mixing with the acid is preferred because it gives a superior distribution. The recommended rate of application is 2.5 oz of Na_2MoO_4/acre of herbage which is supplied by 3 cwt of molybdenized superphosphate. They emphasize that it is important to maintain a proper balance between molybdenum and copper in the herbage, and to prevent a copper deficiency they apply Mo about once in three years unless competent advice recommends otherwise.

Some Recent Reports of Experiments Involving Mo

GEORGIA

(1) Application of Na_2MoO_4 at the rate of 8 oz/acre increased average alfalfa yields from 2,442 lb/acre, oven-dry forage, to 5,256 lb/acre in a three year test by the Georgia AES. The most outstanding effects of the Mo treatment occurred on test plots with "low" lime treatments of 500 lb/acre; on these plots the yield was increased by 2,814 lb/acre. On "high" lime treated plots of 4,000 lb/acre the Mo treatment increased the yield by an average of 1,729 lb/acre. The researchers said that an apparent function of the Mo is to favor nitrogen fixation

in the soil, because the Mo treated alfalfa was greener and assayed a higher nitrogen content than untreated alfalfa (U.S.D.A. Office of Information, April, 1960).

(2) The Georgia AES reported Mo boosted soybean yields by as much as 55% in their field trials. In the tests a seed treatment of $\frac{1}{2}$ lb Na_2MoO_4 per acre of soybeans, grown in acid soil which was not limed, increased yields by 16.7 bu or 55%/acre. The same amount of Mo applied as foliar spray increased yields by 14 bu of 46%/acre. Although lime increased the yield equally as well as the Mo, no increase was obtained from applying lime above 2 tons/acre. The experimental evidence indicates that the use of Mo where soil pH is below 6.0 will benefit soybeans.

WASHINGTON

Soil pH was found to be of greatest importance in the availability of soil Mo to plants.

Application of fertilizers containing sulfate salts tend to depress the uptake of Mo from acid soils. In trials conducted on Palouse soil an application of 1 lb/acre of Mo with 180 lb of sulfur was not sufficient to produce normal plants, whereas the same amount of Mo applied alone was excessive. The plants on plots receiving heavy applications of sulfur alone were extremely deficient in Mo.

On some soils reponse to Mo is nil if lime has been applied; on soils very low in available Ca responses occur only when lime has been applied, and on other soils lime may have little or no effect on the response to Mo. Conversely, on some soils large responses to lime are obtained only when supplemental Mo is supplied.

The effects of this interaction are illustrated by the data in Table 10.1.

AUSTRALIA: WHEAT NITRATE AND MOLYBDENUM[28]

Some Australian red-brown soils used for pot tests for wheat in 1962 produced symptoms of damage such as tip scorch, tiller death, and reduced growth four weeks after germination. In 1964 other soils in pot tests growing wheat at four levels of N supplied as nitrate with or without Mo (as Na_2MoO_4) produced similar symptoms without the Mo additions, but did not do so with Mo. Damage in the 1962 series

[28] Freney, J. R., and Lipsett, J., *Nature*, **205**, 616 (1965).

**TABLE 10.1   EFFECT OF Mo AND LIME ON YIELD,
N AND Mo CONTENTS OF PEAS**

| TREATMENT* | LASHAW FIELD | | | HOUSTON FIELD | | |
| | SEED | PLANT TOP | | SEED | PLANT TOP | |
	YIELD lb/acre	N %	Mo ppm	YIELD lb/acre	N %	Mo ppm
None	770	2.9	0.17	830	2.0	0.16
0.4 Mo	1,190	3.0	0.23	980	2.5	0.16
Lime	1,290	3.0	0.21	1,090	2.3	0.12
Lime + Mo	1,090	3.0	0.78	1,200	2.8	0.60

* Treatments were applied prior to preparation of seed bed. Mo was applied as Na_2MoO_4 and the lime as $Ca(OH)_2$ in amounts to raise the pH level to 6.5.
Source: Wash. State College, Pullman (1956).

could have been due to excessive nitrate supplied; in the 1964 series nitrate, as applied, exhibited yield reducing effects when applied at more than 80 ppm (which seemed the rate for maximal response without damage); however, when Mo was also supplied, higher rates of nitrate than 80 ppm were not associated with symptoms of plant damage though yields did not rise. It was concluded that although the mechanism of yield depression is not known, this condition is alleviated by Mo. To explain the effects of Mo they are being further investigated to see whether (1) the high NO_3 supply reduces the Mo uptake by wheat plants; and (2) the accumulation of nitrate in the wheat plants causes a bigger demand for nitrate reductase enzyme, of which Mo is known to be a component.

With an increasing use of nitrate fertilizers in the world, perhaps the need to pay more attention to molybdenum-nitrate interaction is shown by these and other test data.

General References

(1) Borys, M. W., and Childers, N. F., "The Role of Molybdenum in Plants and Soils," Rutgers University, New Brunswick, New Jersey, 1960.

(2) Allcroft, R., and Lewis, G., "Relationship of copper, molybdenum and inorganic sulfate contents of feeding stuffs to the occurrence of copper deficiency in sheep and cattle," *Landbouwk. Tijdschr., Wageningen,* **68**, 711, 23 (1956).

(3) Dick, A. T., "Molybdenum and Copper Relationships in Animal Nutrition," ed., McElroy, W. D., and Glass, B., "Inorganic Nitrogen Metabolism," Baltimore, Johns Hopkins Press, 1956.

(4) Stiles, W., "Trace Elements in Plants," 3rd. edition, Cambridge Press, 1961.

11

COPPER

Copper is an essential plant-animal micronutrient. Its essentiality was first conclusively proved by the experimental work conducted by Anna L. Sommer[1], in which the especially purified chemical compounds and the water redistilled from pyrex equipment for culture solutions were crucial factors. The failure of earlier experiments using the solution culture technique to show a response to copper may be attributed to the traces of copper occurring as an impurity in the highly concentrated fertilizer salts and in the distilled water produced from copper stills.

Copper deficiency in plants was first recognized as "reclamation disease" in crops grown on newly reclaimed peats in Holland and later in crops grown on some sandy and gravelly soils in Florida and Australia. J. Hudig and C. Meyer discovered in 1925 and 1926[2] that this disease could be prevented for a period of years by applying one dressing of 50–100 lb of copper sulfate ($CuSO_4 \cdot 5H_2O$) per hectare.

---

[1] Sommer, A. L., "Elements necessary in small amounts for plant growth," *Am. Fert.*, **72**, (6), 15–18 (1930).
"Copper as an essential to plant growth," *Plant Physiol.*, **6**, 339–345 (1931).
[2] Hudig, J., and Meyer, C., "De ontginningsziekte en haar bestrijding," (De Veldbode), (1925, 1926).

This result was subsequently confirmed through a strictly controlled experiment using the *Aspergillus niger* bioassay technique.

In a comparison in 1933, E. Brendenburg established the identity of the symptoms of diseased oat plants grown in newly cultivated peat soils with those caused by copper deficiencies induced in solution cultures[3]. Allison, et al., (1927) were able to produce crops on certain otherwise unproductive peats of Florida by the application of copper sulfate[4].

Copper seems to have two functions, each quite separate from the other. One, concerned with plant nutrition, is to increase oxidase activity of the ascorbic acid enzyme and to influence other metabolic reactions. For example, Hill reported that copper is essential for the formation of iron porphyrin, precursor of chlorophyll, even though copper is not a part of the chlorophyll molecule[5]. The other function is concerned with the condition of the soil, previously noted in connection with peaty soils. W. S. Smith thought the effect of copper in copper deficient peat soils was to precipitate or inactivate certain toxins present in the peat[6].

Copper Chemistry

Copper has an atomic number of 29 and atomic weight of 63.54. Its chemical symbol is Cu. In the Periodic Table it is classed as a transition element and forms a subgroup with silver and gold. The members of the copper subgroup, copper, silver, and gold, are among the oldest known elements, since they are frequently found in nature in the uncombined or native state. Because of their relative scarcity and resistance to corrosion, man used them originally for decorative purposes and in time adapted them for use in coins. It is surprising that, although copper is a very familiar metal with many uses, it makes up only 0.0001% of the earth's crust.

Its deposits are concentrated and economically worked. The element in its native state occurs as 99.9% pure and in the form of two principal classes of minerals: sulfide ores and oxide ores. The principal

[3] Brendenburg, E., "Researches on reclamation diseases," *Tijdsch. Plantenziekte,* **39**, 189–192 (1933).
[4] Allison, R. V., et al., *Florida Agric. Expt. Sta. Bulletin,* 190 (1937).
[5] Hill, R., Brit. C. W. Sci. Office Cong. Sess. B., Plant Nutrition, Australia (1949).
[6] Smith, W S., Dissertation, Wageningen (1929)

sulfide ores are chalcosite (Cu_2S), chalcopyrite or copper pyrites ($CuFeS_2$), and covellite (CuS); the principal oxide ores are cuprite (Cu_2O), malachite ($CuCO_3 \cdot Cu(OH)_2$), and tenorite (CuO). About 80% of current copper production is from the sulfide ores.

The most familiar copper compound is copper sulfate penta-hydrate ($CuSO_4 \cdot 5H_2O$), also designated as $Cu(H_2O)_4\ SO_4 \cdot H_2O$ and commonly called blue stone or blue vitriol. This compound is used extensively as a fungicide and germicide, since the cupric ion (Cu^{2+}) is toxic to lower forms of life, as for example, to algae, the scum that forms on water supplies. It has been recognized for more than 100 years as a fungicide. Bordeaux mixture (copper sulfate and lime), first used by Millardet as a fungicide in vineyards near Bordeaux, France about 1885, still remains one of the more effective fungicides applied as a dormant spray. Some plants sensitive to copper show fruit russeting from Bordeaux, and this led to the finding of substitutes such as the fixed (insoluble) coppers; they include basic copper sulfate, copper oxychlorides, copper silicate, cuprous oxide, and mixtures of these. Both the cupric (Cu^{2+}) and the cuprous (Cu^{+}) forms of copper com-pounds are used as pesticides.

The fixed coppers are used largely for treating vegetables but are also used in certain orchard spray programs. Basic copper sulfates vary in their formula, owing to the ratio of the ingredients and to the com-bining compounds. The carbonates are preferred to hydroxides for such combinations.

TABLE 11.1   COPPER COMPOUNDS AND Cu CONTENT

COMPOUND	CHEMICAL FORMULA	% Cu FROM FORMULA
Copper pyrite	$CuFeS_2$	34.6
Bornite with copper glance	$CuFeS_2$	63.3
Chalcosite	CuS_2	79.9
Tenorite	CuO	75
Malachite	$CuCO_3 \cdot Cu(OH)_2$	57.5
Azorite	$2CuCO_3 \cdot Cu(OH)_2$	55
Cuprite	Cu_2O	88.8
Copper sulfate, pentahydrate	$CuSO_4 \cdot 5H_2O$	25.5
Copper sulfate, monohydrate	$CuSO_4 \cdot H_2O$	35

For making suitable fungicides investigators experimented with calcium carbonate ($CaCO_3$), basic magnesium carbonate ($4MgCO_3 \cdot Mg(OH)_2 \cdot 5H_2O$), and sodium carbonate (Na_2CO_3), to combine with copper sulfate in amounts sufficient to cause complete precipitation. The basic sulfates are marketed under different names including Tribasic, Basicap, Coppatone, Copper hydroxysulfate. Other copper products include copper ammonium zeolite, called Z-O, copper carbonate, copper phosphate, copper ammonium fluoride, copper acetate, copper naphthenate, and cuprous cyanide. These several compounds differ in copper content, but each has been developed either as a fungicide or pesticide to meet certain conditions of soils and plants.

Table 11.1 lists the common copper compounds and their copper content:

Solubility of Copper Compounds in 100 Parts Water

Copper sulfate ($CuSO_4 \cdot 5H_2O$) . . . . .21.97 at room temperatures
206 at 100°C
Basic copper carbonates. . . . . .only a trace
Basic copper sulfate . . . . .only a trace

Copper in Soils

Copper deficiency frequently occurs in soils of high organic content, particularly in newly cultivated peats and in alkaline peat soils which have appreciable amounts of ferrous iron. Sandy and open gravelly soils also lack sufficient available quantities. Copper placed in organic soils is held tightly in the zone of placement ; similarly in clay loam soils having an appreciable amount of organic matter. In one test an organic soil was treated with 50 lb of $CuSO_4 \cdot 5H_2O$/acre, but after a period of 5 years an approximate equivalent of 48 lb of copper sulfate was found in the upper 8 in. This experience is referred to by soil scientists to explain why shallow rooted crops seem to respond better to copper treatment than the deep rooted types. Another characteristic of organic and heavy clay soils is that they impede easy leaching of soluble copper, "fixing" the copper, making it less available. Such soils require a heavier dosage of copper to correct a deficiency than the more open lighter soils.

The amount of copper in soils varies; a normal agricultural soil is estimated to contain from 1 to over 50 ppm. The Cu cation is adsorbed strongly by the negatively charged colloidal complex of clay soils but is easily displaced by other cations. If the sulfur anion is also present in the soil, copper will tend to react with it to be held as an almost insoluble copper sulfide.

Mitchell[7] studied the distribution of trace elements in the soils of northeast Scotland and found by spectrographic means that their copper constituents ranged in quantity from 5–5,000 ppm. He emphasized, however, that the determination of total amount is not too significant, since only that portion which is available really counts. Perhaps the total may serve as a guide to the possibility that a deficiency or an excess may be occurring.

Despite the lack of available copper in normal, highly organic soils, as for example, the saw grass peats of Florida, experience has shown that such unproductive soils can be made to grow profitable crops by a judicious application of sizeable amounts of copper, usually as the copper sulfate pentahydrate compound. Truck farmers on the Carolina's coastal plain soils customarily add 200 lb/acre of $CuSO_4 \cdot 5H_2O$ to newly cleared peat soils followed by about 50 lb/acre per year thereafter.

R. S. Holmes[8] studied the contents of a large number of United States soils and reported the copper content as ranging on average from 2–67 ppm. The old, leached acid soils of the Southeastern States averaged lower in total copper content than the soils of the Southwest. T. S. Brun[9] reporting on his studies of humic soils in Norway indicated that their copper content could be divided into three groups: water-soluble, absorbed, and fixed. The fixed copper could not be released or separated without the chemical decomposition of the soil, and therefore could not be considered an available source to plants. He also emphasized that a soil could have a high total copper content but a low available amount.

The factors generally recognized as directly influencing the soil's content of available copper are; pH value, humus content, the nature of the previous crop, the proportion of sand to clay, the presence of other chemical elements with which the copper could interact. It is possible for several of these factors to be active at the same time. For

[7] Mitchell, R. L., *Proc. Nutr. Soc.*, **1**, 183–189 (1944).
[8] Holmes, R. S., *Soil Sci.*, **56**, 359–370 (1943).
[9] Brun, T. S., Bergens Museum Arbok, Naturv. Rekke, 1945.

example, copper could react with sulfur and form an insoluble CuS_2, or react synergistically with iron to form iron porphyrin, or form complexes with lactates and tartrates. It is known that the greater the acidity of a peat, the greater may be the relative response of a crop to added nutrient copper, and the larger the number of crops likely to benefit from the applied copper[10].

F. A. Gilbert[11] investigated the culture of tobacco, using over 200 field tests, in an effort to determine the value of applied nutrient copper on soils that were not actually deficient in available copper. He grew Burley tobacco in four Ohio Valley states, flue-cured tobacco in Virginia and the Carolinas. The available copper ranged from 1 ppm in a few Coastal Plain soils to 16 ppm in some of the Ohio valley soils, with the majority having, on average, about 5–10 ppm. About 80% of the copper treated plots responded to the treatment, the greatest increase in crop production was 35%, and the average increase for the treatments was 8%.

J. E. McMurtrey[12], a tobacco specialist, emphasizes that the amount of copper needed by plants is very small. In his tests using nutrient solutions, he found that the amount needed to correct a deficiency was from $\frac{1}{16}$–$\frac{1}{8}$ ppm., equivalent to $\frac{1}{2}$ oz of copper in about 30,000–60,000 gallons of water; an amount in excess of these produced definite stunting of the test tobacco plants.

T. Wallace observed[13] that in the neighborhood of copper mineral deposits the soils are generally toxic to plants. The amount of copper in normal agricultural soils is very small and often is there as the result of concentrations and additions from growing plants and some added residues.

Peat soils tend to adsorb or fix soluble copper and to tenaciously hold on to it. For this reason experience dictates that applied copper sulfate, the form almost exclusively applied, be of the small crystal kind rather than the pulverant type; besides, the crystal form lasts longer and the danger of releasing an excessive amount is lessened.

Experimental evidence shows that the $CuSO_4 \cdot 5H_2O$ commonly employed as a soil amendment is not necessarily the only carrier for this purpose; finely ground copper minerals of which several contain

[10] Harmer, P. M., *Soil Sci. Soc. Am. Proc.*, **10**, 284–294 (1945).
[11] Gilbert, F. A., *Better Crops Plant Food*, **32**, 8–11, 44–46 (1948).
[12] McMurtrey, J. E., "Hunger Signs in Plants," 3rd Ed., Nat'l Plant Food Inst., 99–142, 1964.
[13] Wallace, T., "Diagnosis of Mineral Deficiencies in Plants, H.M.S.O., 1943.

more of the copper metal can also be effectively used. Included among these other compounds are pyrites ash, malachite, copper flake, cuprous and cupric oxides, copper acetate, copper carbonate, several copper-bearing sulfide ores, and chalcopyrite. Pyrite ash, with 1–2% Cu, obtained as a by-product when pyrite is roasted to produce sulfuric acid, has been used as a copper soil amendment in Denmark. Australian investigators have compared oxidized copper ore, roaster residues from pyrites burners, and $CuSO_4 \cdot 5H_2O$ as sources of copper for fertilizer purposes and found all three to be effective, but no one form superior to the others under all conditions. Some doubt existed regarding the fairness of the test insofar as it concerned the copper sulfate used, because it alone did not contain zinc as an impurity and some evidence existed that some of the test soils were deficient in zinc[14]. In the United States comparisons were made between copper flake, copper oxides, and other copper compounds with $CuSO_4 \cdot 5H_2O$, and the result was that although these other sources compared favorably with the copper sulfate, none surpassed it in effectiveness. Fertilizer manufacturers who plan to include a copper carrier in the formula are advised to consider that a high Cu concentration in the carrier is important from the viewpoint that the NPK of the mixture will be less diluted by the addition of the copper carrier, and furthermore, the less soluble forms are less likely to release toxic amounts of metallic copper during the plant's early growth period.

Copper deficiency has figured prominently in Australian agriculture. Consider the case of the Ninety Mile Plain of South Australia, a vast waste land on which only stunted scrub of heath and eucalyptus thickets would grow. Soil research revealed that the area's soil lacked copper and zinc, and that by adding copper and zinc compounds with superphosphate, the soil could be converted into productive land capable of growing a lush pasture suitable for sheep grazing. An area of 6,000,000 acres was thus reclaimed and is now divided into farms in which two sheep per acre are easily sustained where previously one sheep required at least 20 acres.

New Zealanders have also found copper deficiency on their farm lands. It is a common custom there to add trace elements in the fertilizer mixtures; finely crystalline blue vitriol is dry mixed at the rate of 2.5% of the solid fertilizer. Currently the local agriculture consumes annually about 20,000 tons of this copperized fertilizer

[14] Teakle, L. J. H., et al., *J. Dep. Agr. W. Australia*, **18**, 70–86 (1941).

mixture. The copper is used on peat soils to satisfy crop needs, and additional copper is employed for maintaining the health of the livestock. The authorities insist that copper and molybdenum be kept in balance in the forage after it was disclosed that applications of molybdenum on some pastures induced a deficiency of copper. It is a known fact now that copper helps regulate the availability of molybdenum in pastures by preventing accumulations of Mo to reach toxic proportions.

Fertilizer manufacturers are understandably reluctant to include copper materials in their general fertilizer mixtures, not only because of the added cost but primarily from fear regarding the potential toxicity carried over a number of years. Gilbert states[11] that in some of his field experiments the growth of tomatoes in nutrient solution was depressed by as little as 1 ppm Cu, and Teakle reported[14] that toxicity occurred on some very sandy soils of Western Australia from an application of 10 lb Cu/acre. Peat soils, on the other hand, will absorb heavy applications of copper without fear of toxicity to crop plants, in fact, the only fear is that not enough copper will have been applied for maximal effectiveness. Gilbert also affirmed his belief, after a long experience with this phase of crop growth, that copper toxicity is not likely to occur in ordinary farm soils from residual accumulations derived from copper containing mixed fertilizers, even if the mixture should contain as much as 1% of the element.

Nevertheless, the fertilizer industry, aware of the need for copper in the soils of many areas, has been adding copper to many of its fertilizer mixtures. A commonly used fertilizer in Florida would contain, for example, magnesia (MgO), manganese oxide (MnO), and CuO, in addition to the regular N, P_2O_5, and K_2O ingredients.

A survey made in 1962 to determine copper deficient areas in the United States revealed that it was reported in 13 states. K. C. Berger found copper deficient in Wisconsin even on some upland mineral soils[15]. He records that cereal grains, corn, alfalfa, and some vegetable crops have a high requirement for copper. In a test in which oats were grown on an acid peat soil deficient in Cu, Berger applied 9 tons limestone, 700 lb of 0-10-30, 40 lb $MnSO_4$, 20 lb $ZnSO_4 \cdot 7H_2O$, and 5 tons cow manure to the acre. The yield of oats was practically zero. When he added 20 lb of CuO/acre the oat yield jumped to 74 bu/acre. This illustrates the essential character of copper in growing crops on peat soils.

[15] Berger, K. C., "Introductory Soils," The Macmillan Co., New York, 1965.

Copper in Enzymes

Copper was among the earliest micronutrients to become thus identified. It is an essential nutrient for plants and animals; it is known to be associated metabolically with a number of metalloproteins, and it has been demonstrated to be deficient in certain types of soil. Copper plays an essential role in the respiration of many of the lower forms of animal life through the hemocyanins, the copper containing respiratory proteins found in the blood of certain marine animals. These copper protein enzymes display a remarkable specificity, as for example ascorbic acid oxidase and tyrosinase; these two copper proteins have the same copper content and protein properties, but have a definitely specific enzyme activity[16].

Copper enzymes have aroused the interest of biochemists for many years. It was G. Bertrand who first designated the copper catalytic agent we know as tyrosinase[17]. True tyrosinases occur widely in nature, often in joint activity with other enzymes. The enzymes are found principally in plants and invertebrates and less frequently among the higher animals. Failure to recognize early in research that the enzyme tyrosinase contains essential copper was due largely to difficulties in preparation. The plants containing the tyrosinase oxidases include many fruits and vegetables which darken upon being injured, for example, tea, mushrooms, beans, wheat, artichokes, fungi of several genera, spinach, apple leaves, and certain bacteria. Lower animals in which these enzymes have been reported include cuttlefish, certain crustacea, and a number of insects and larvae. In short, the tyrosinases comprise one of the most extensively distributed groups of enzymes in nature[18].

Mulder[19] found that an external copper supply had a great influence on the tyrosinase activity of potato tubers; grown in soils poor in copper, potato tubers had a tyrosinase activity of less than 0.1 than in tubers grown in comparable plots fertilized with copper sulfate. The low tyrosinase activity was manifest in only slight blackening of bruised tubers deficient in copper compared with tubers containing a normal amount.

[16] Dawson, Charles R., "Copper Metabolism Symposium," Johns Hopkins Press, 18–47, 1950.
[17] Bertrand, G., Sur une nouvelle oxydase, C.r. Acad. Sci., Paris, **122**, 1215 (1896).
[18] Mallette, M. F., Copper Metabolism Symposium, Johns Hopkins Press, 48–75, 1950.
[19] Mulder, E. G., *Plant Soil*, **2**, 59–121 (1949).

Copper is an important constituent of two other enzymes, namely, ascorbic acid oxidase and laccase; the oxidase catalyzes the oxidation of ascorbic acid (vitamin C) in the presence of oxygen, and laccase oxidizes phenolic compounds but not tyrosine[20].

Copper in Green Plants

That plants need copper for their normal nutrition was not discovered as an isolated fact. Prior to its establishment as an essential plant nutrient, it was accepted in scientific circles that plants required as essential elements carbon, hydrogen, oxygen, nitrogen, potassium, phosphorus, magnesium, sulfur, calcium, and iron. At first when it was reported that copper increased the yields of green plants, the effect was ascribed to chemical stimulation. Many references in the literature had referred to the stimulating action of Bordeaux fungicide, a lime-copper sulfate spray. Reports were current that copper stimulated the growth of onions on the peat soils of Florida and other crops grown on muck soils. The introduction of the effective calcium carbonate method of purification of nutrient solutions by Robert A. Steinberg of the U.S. Dept. Agr. in 1919[21], made it possible to demonstrate conclusively that the theory of chemical stimulation had no basis in fact. Steinberg showed that impurities in trace elements could be readily and effectively removed by adding calcium carbonate to the neutral medium, heating, and then filtering. Precipitation of the impurities as phosphates and carbonates, and their adsorption on the precipitate of dicalcium phosphate results in an almost complete removal of trace element impurities.

In 1927 Bortels published the results of his research on iron, zinc, and copper as a nutrient for microorganisms[22]. Four years later 2 papers appeared almost simultaneously in which the authors also claimed the essentiality of these same elements for green plants; Miss A. L. Sommer[23] grew the tomato plant, sunflower, and flax in water culture, which demonstrated a decline of more than 90% in yield when copper was not added, and Lipman and MacKinney[24] growing barley

[20] Stotz, E. H., et al., *Science*, **86**, 35 (1937).
[21] Steinberg, R. A., *J. Agr. Res.*, **51**, 413–424 (1935).
[22] Bortels, H., *Biochem. Fert.*, **182**, 301–358 (1927).
[23] Sommer, A. L., *Plant Physiol.*, **6**, 339–345 (1931).
[24] Lipman, C. B., and MacKinney, G., *Plant Physiol.*, **6**, 593–599 (1931).

in water culture found it was unable to develop seed when copper was omitted and not until $\frac{1}{16} - \frac{1}{8}$ ppm of copper was supplied was the disability corrected. These claims were soon substantiated by research using tobacco, oats, oranges, and many other crops. The accumulation of evidence made it clear that copper in minute quantity is essential for the growth and reproduction of green plants. Copper applications in the production of commercial crops has become the practice not only for fertilization but also for fungicidal purposes.

Fertilization with copper may be for correcting abnormalities in plants or for meeting a nutritional need in normal plants. The prophylactic purpose may be for correcting reclamation disease on newly cultivated peat soils or for correcting dieback disease of citrus trees.

Lucas studied the growth of various crops grown on muck soils[25]. In Table 11.2 following, as prepared by him, it is shown that only three out of sixteen crops grown on muck soils failed to respond to application of copper.

Copper occurs throughout the tissues of plants but is concentrated mostly in the green leaves and in the germ of seeds. Green[26] has reported that plant tissues on average contain from 3–40 ppm depending on species, soil, amount of fertilizer used, and other factors. The amount of copper in plant tissues can be increased by fertilization; in a sand culture Gilbert[27] increased the content of copper of corn ear from 5–28 ppm.

E. G. Mulder demonstrated with plants grown in different soils how copper influences the correction of reclamation disease. The following summary, Table 11.3, was prepared by him. Wheat was the test plant which was grown in normal and deficient (reclamation) soils. He analyzed the grain and straw of mature plants, and leaf and stem material of young plants. The results proved that reclamation disease resulted from the absence of copper in the soil.

Deficiency Symptoms

It is rare to find copper deficiency in tobacco culture under field conditions. The fixing power of the soil influences the need for application

[25] Lucas, R. E., *Soil Sci.*, **65**, 461–469 (1948).
[26] Green, H. H., *Vet. Rec.*, **50**, 37, 1185 (1938).
[27] Gilbert, F. A., *Better Crops Plant Food*, **32**, 8–11, 44–46 (1948).

**TABLE 11.2   EFFECT OF COPPER FERTILIZATION ON THE
Cu CONTENT
OF SPECIFIED PLANTS (DRY WEIGHT BASIS)**

| CROP | RESPONSE | COPPER IN PLANTS, ppm | |
		NO COPPER	WITH COPPER*
Alfalfa	Fair	5	10
Barley	Good	10	14
Carrot roots	Good	3	5
Dill seed	Good	6	12
Head lettuce	Good	3	9
Ladino clover	None	7	14
Oats (preheading)	Good	11	15
Onion bulbs	Fair	2	5
Peppermint	None	8	12
Red clover	None	7	15
Sudan grass	Good	5	10
Sugar beet tops	Good	6	7
Spinach	Good	8	12
Tomato fruit	Fair	4	8
Wheat grain	Good	6	8
Wheat (preheading)	Good	8	12

* 100 lb copper sulfate/acre.

of fertilizing copper. The chemical properties of the soil will suggest the advisability of including a copper compound in the fertilizer mixture, unless it is an established fact that copper is definitely lacking. Perhaps it is preferable to apply the copper in a spray directly on the plant, since it is known the copper cation will combine with phosphate ion to form insoluble cupric phosphate.

The color of legume and forage plants deficient in copper tends to be grayish-green, blue-green, or olive green. Alfalfa leaves turn to a palish green with a grayish cast; the plant shows a stunted growth; the internodes become shortened to produce a bushy type of plant.

Deficiency symptoms in corn (maize) plants (see Figures 11.2, 11.3) appear first on the youngest leaves, more frequently on immature plants. The early symptoms are evident in the yellowing of the upper or youngest leaves and a slight stunting in growth. In severe deficiency the plant is definitely stunted in its growth, the leaf tips curl, the younger leaves become very pale yellow and some of the older leaves

**TABLE 11.3   COPPER CONTENT, mg/kg DRY WEIGHT OF
WHEAT (GRAIN) GROWN ON SPECIFIED SOILS**

SOIL	TREATMENT	REMARKS	Cu mg/kg DRY WEIGHT
Sandy soil	Control	Slightly diseased plants	0.9
Sandy soil	50 mg $CuSO_4$/2 kg soil	Normal plants	3.0
Peaty soil	Control	Slightly diseased plants	0.7
Peaty soil	50 mg $CuSO_4$/2 kg soil	Normal plants	1.3
Peaty soil	Control	Slightly diseased plants	0.9
Peaty soil	100 kg $CuSO_4$/hectare	Normal plants	1.6
Peaty soil	Control	Normal plants	1.2
Clay soil Groningen	Control	Normal plants	6.0
Clay soil Zeeland	Control	Normal plants	2.5

Source: Mulder, E. G., Agr. Expt. Sta., Groningen, Holland.

dieback. Finally, the tips and leaf edges discolor very much, as in potassium deficiency symptoms. Severe cases of deficiency rarely complete their growth cycle and usually die during the early or middle part of the growing season.

Copper deficiency in citrus (see Figure 11.1) is easily recognized as dieback or exanthema, which is characterized by the death of new growth, the formation of many side branches or witches'-broom below the dead portions; gum pockets develop between the bark and wood of the affected branches, and the fruit shows brown excrescences. Usually an application of 50–100 lb copper sulfate/acre in the soil or as a spray will correct the deficiency.

Tomato plants growing in deficient soil are dwarfed, the leaf edges roll inward, and the plant develops a bluish appearance.

Onions suffering from a copper deficiency in the soil show pale yellow bulbs which lack firmness in texture. Most of the plants will contain less than 10 ppm on a dry basis and normally when the Cu content exceeds 30 ppm show toxic symptoms. The affected bulbs develop abnormally thin, yellow scales. On peat soils 100–300 lb copper sulfate/acre increased thickness of scales and changed color to a brilliant brown.

Generally, copper deficiency is limited to those grown on dark colored soils and on peat and muck. Experience on western New York

11.1   Copper deficiency in oranges. (*Courtesy of Fla. Agric. Expt. Station*)

peat soils has been that it was necessary to add 100–200 lb of $CuSO_4 \cdot 5H_2O$/acre in order to produce acceptable vegetables. Florida peat soils require at least 20–30 lb/acre of the copper sulfate to prevent development of deficiency symptoms of most vegetable crops.

Deficiency symptoms of sugar cane plants are as follows: poor development of cane stools, droopy tops, yellowing of leaves, failure of spindles to unroll. The chlorotic condition of the leaves is revealed in a pronounced striping effect, the leaves are softer to the touch, and stalks and spindles have a rubbery consistency and can easily be bent without snapping. The symptoms occur only when the Cu content of the cane leaves falls to about 3–4 ppm or below. Australian growers report correction of the deficiency by applying 50 lb copper sulfate/acre.

11.2 Response to copper treatment is evident in this corn crop. No copper on right; 20 pounds per acre copper sulfate on left. (*Courtesy of National Plant Food Institute*)

Interrelationships

Plant physiologists and soil scientists have studied the interactions between copper and other trace elements as revealed by effects on plant growth. The concept of balanced nutrient solutions and ion antagonism has been developed, whereby if toxicity to the plants is to be prevented the ratios of ions must be within definite limits. The ratios between different ions would be quite different. An early observation noted that plants growing in soils treated with copper might develop symptoms associated with iron chlorosis. One investigator[28] reported symptoms of "gray speck" disease due to deficiency of manganese, when an excessive dose of copper was added to solution or sand cultures. Copper and aluminum are reported as being antagonistic. Excesses of zinc, copper, manganese, cobalt, or nickel are reported as producing symptoms in flax very closely resembling those of iron chlorosis. In other words they were antagonistic to the element

[28] Mulder, E. G., *Z. Pflanzenbau Pflanzenschutz*, **50**, 230–264 (1940).

11.3 Ragged edges of leaves and short internodes of corn plants is an indication of severe copper deficiency. (*Courtesy of National Plant Food Institute*)

iron. The addition of molybdenum apparently reduced all the symptoms of toxicity caused by these several trace elements.

Copper may prevent the excessive absorption of a second element that is toxic if present in too great an amount, or it may increase the availability of the second element. Copper may affect or be affected by several other elements; a good example is the copper-nitrogen relationship. Where copper deficiency occurs, an increase in the nitrogen level will intensify the severity of the deficiency symptoms. A study of the interrelation between available nitrogen and copper in producing dieback in the tung trees showed that high nitrogen aggravated the copper deficiency as it does in exanthema of citrus trees.

Another example is the relation of copper and zinc; the addition of zinc sulfate on organic soil favored plant growth only when copper was present.

The relationship of copper to molybdenum has previously been noted; in the disease known as "teartness" or "scouring" affecting sheep and cattle grazing pasture containing a high level of molybdenum, copper has proved to be a specific remedy for the disease.

Copper in Animals

Copper is also an indispensable element in the nutrition of animals and humans. Analyses of the organs of man and many species of animals for determining the content of copper are in the published records. In general, the tissues of the brain, kidney, liver, and heart contained the most copper, while the skin, lungs, pancreas, spleen, and flesh had the least. In all species studied, it was found that the adult liver had the greater amount of copper than the other adult organs, but much less than the livers of newly born animals. A high storage of copper in the young liver is necessary to provide the young animal with a supply of copper over the suckling period when only insignificant amounts of this element are available in the diet of milk. Animals are known to suffer anemia when confined to a diet made up solely of milk, because the missing copper plays an important role in the metabolism of iron. Cow's milk produced on a normal ration contains about 0.15 mg of copper per liter (ppm). Another estimate is that uncontaminated cow's milk contains about 0.5–0.2 ppm and human milk 0.5–0.6 ppm. The copper content of an average egg is about 0.07 mg, concentrated largely in the yolk.

Functions of Cu

In the utilization of iron for the formation of haemoglobin, copper plays an important part. Without copper iron is assimilated and stored in the liver but is not converted into haemoglobin. The animal body contains a relatively minute amount of copper, but a deficiency of this element in the food of the animal or an interference with the utilization of that copper by the animal can lead to serious consequences.

Anemia in young pigs is quite a common ailment. In the preventive treatment of this condition, a small amount of copper is added to the iron used for correcting the condition. The reason is that milk is poor in copper and in the metabolism of iron the presence of copper is necessary.

In Australia a disease occurs among sheep and is named "enzootic ataxia," which most often attacks 1–2 month old lambs. The animal develops a stiff gait, and its growth is retarded. An afflicted lamb rarely survives beyond 3–4 weeks. Australian scientists found that the condition is prevented by drenching ewes with copper sulfate. Research at the University of London has shown that the cause of the disease is the very low copper status of the animal and of the herbage (less than 5 ppm, dry weights).

Another disease reported in the United Kingdom is described as "swayback" or "swingback." It is prevented by giving copper to pregnant ewes. Investigators are inclined to the belief that swayback and ataxia are one and the same.

"Teart"

A considerable pasture area in central Somerset, England, causes scouring in grazing cattle. Such land is locally described as "teart." It is known that the basic cause of the condition is the molybdenum content of the herbage. The remedy is to feed or drench the animal with copper sulfate. A dose of 2 gm of copper sulfate per day for cows and one gram for young stock suffices to cure and prevent the scouring condition.

Other "Diseases"

Other conditions in different countries associated with copper deficiency go by local names, such as "falling disease" of dairy cattle (Australia); "wobbles" of foals and calves (Australia and U.S.A.); "stringy" wool of sheep (Australia); "salt sick" of cattle (Florida) and "coast disease" of grazing animals (Australia) are associated with a dual deficiency, namely copper and cobalt.

In the case of "stringy" wool the sheep's wool loses its crimp as the result of copper deficiency in the animal. The condition is remedied by either applying 5 lb/acre of copper as copperized superphosphate or by providing the animals with copper containing salt licks. If the deficiency is severe, the breeding of ewes and the quality of the wool both suffer. Furthermore, lambs may be stillborn or born with a wobbly gait and die soon after birth.

Calves born to cows on such copper deficient land rarely suffer as severely as lambs, but cows frequently are afflicted with a malady named locally as "falling disease," because the affected cow may suddenly drop dead in the field from heart failure; copper deficiency in the cow weakens the heart muscle to a degree whereby a momentary stress proves fatal. Soil treatment with copper sulfate or copperized superphosphate has successfully eliminated the cause.

Using copper compounds prepared from radio-labelled copper (^{64}Cu), studies have been made of a steer's ability to absorb copper and lose it by excretion. Powdered cupric oxide was slowly absorbed but quickly voided. Copper as needles or wire was slowly absorbed but also slowly lost. Cuprous oxide was absorbed fairly quickly but lost readily. Copper carbonate showed the highest rate of absorption but also the highest rate of loss. Copper nitrate and copper sulfate had similar rapid rates of absorption and fairly slow rates of excretion[29].

Copper as a Fungicide

Copper compounds have been used for almost 100 years as control agents of plant diseases, for example, to control wheat bunt and smut, and at present they are regarded as a principal means of protecting plants against disease. Copper sulphate and lime (Bordeaux mixture) remains the leading fungicide for dormant trees. Some plants are sensitive to copper and show foliage injury and russetting of fruit from Bordeaux. To avoid this a number of substitutes have been developed that are less injurious; the so-called fixed or insoluble coppers qualify. They include basic copper sulfate, copper oxychlorides, copper silicate, cuprous oxide, and mixtures of these. Metallic copper is slightly fungicidal in itself and is used to treat pruning wounds of trees. Both the cuprous and cupric forms of copper compounds are used as pesticides[30].

Comparative Toxicity

Apparently, no copper compound which combines high fungicidal value with low leaf-damaging properties has been developed. In this respect Bordeaux mixture has as good a range as any of the other compounds now employed. The results of many comparative tests during the past 50 years indicate that no copper compound has proved equal to Bordeaux mixture[31].

[29] Chapman, H. L., and Bell, M. C., *J. Animal Sci.*, **22**, 82 (1963).
[30] deOng, E. R., "Chemistry and Uses of Pesticides," New York, Reinhold Publishing Corp., 1956.
[31] McCallan, S. E. A., and Wilcoxon, F., *Contrib. Boyce Thompson Inst.*, 9, 249 (1938).

Since 1934 plant pathologists have shown considerable interest in discovering organic fungicides like the patented dithiocarbamates, a group of fungicides used effectively as seed disinfectants, orchard sprays, and insect repellents. The purpose was to find safer and more effective chemicals by combining the fungitoxicity of copper with that of organic radicals. Thus, copper resinate, copper acetate, copper oleate, copper naphthenate, and other copper salts of organic acids came into being. Their commercial value, however, has been limited to special uses as, for example, wood preservatives and garden fungicides. One of the latest of this type has been announced under the trade name, TC-90, formulated with 48% copper salts of fatty and rosin acids (metallic Cu, 4%), and with a petroleum distillate and an emulsifier as the inert ingredients. TC-90 is recommended for use on peanuts, cucurbits, and citrus, and for concentrate or dilute spraying.

Continued intake of small amounts of copper, usually copper sulfate, is known to have poisoned sheep and other livestock. This form of injury is reported from Texas where medicated copper mixtures are given continuously throughout the year to sheep to remove intestinal worms (*Texas Agr. Exp. Sta. Bull.*, 499 (1934)).

Some General References

Arnon, D. I., *Plant Physiol.*, **24**, 1–15 (1949).

Arnon, D. I., and Stout, P. R., *Plant Physiol.*, **14**, 371–375 (1939).

Beeson, K. C., *U.S. Dept. Agr. Misc. Publi.*, No. 369 (1941).

Berger, K. C., "Introductory Soils," New York, The Macmillan Co., 1965.

Cunningham, I. J., *New Zealand J. Agr.*, **69**, 559–569.

Cunningham, I. J., *New Zealand J. Agr.*, **72**, 261.

Cunningham, I. J., *New Zealand J. Sci. Technol.*, **27A**, 372–376, 381–396 (1946).

Gilbert, F. A., *Better Crops Plant Food*, **32**, 8–11, 44–46 (1948).

Gilbert, F. A., *Advan. Agron.*, **IV**, 147–173, Academic Press, N.Y. (1952).

Harmer, P. M., *Soil Sci. Soc. Am. Proc.*, **10**, 284–294 (1946).

Holmes, R. S., *Soil Sci.*, **56**, 359–370 (1943).

McElroy, W. D., and Glass, B., "Copper Metabolism Symposium," Maryland, Johns Hopkins Press, Balto., 1950.

McMurtrey, J. E., "Hunger Signs in Plants," Washington, D.C., Nat'l Plant Food Inst., 1964.

Mitchell, R. L., "Trace Elements in Chemistry of the Soil," New York, Reinhold Publishing Corp., 1955.

Sommer, Anna L., "Copper and Plant Growth," *Soil Sci.*, **60**, 71–79 (1945).

Stiles, W., "Trace Elements in Plants," 3rd Edition, Cambridge University Press, 1961.

Swaine, D. J., "Trace Contents of Soils," *Commonwealth Bur. Soil Sci., Tech. Commun.*, No. 48 (1962).

Tisdale, W. B., *Fla. Agric. Exp. Sta. Report*, No. 135 (1930).

Wallace, T., "Diagnosis of Mineral Deficiencies in Plants," H.M.S.O. (1943).

12

CHLORINE

Chlorine (Cl) is the latest element to be established and accepted as a micronutrient essential to both plant and animal life, its essentiality being positively demonstrated in 1954 by T. C. Broyer and three associates[1]. Prior to this date chlorine had been regarded as being beneficial but not indispensable for normal plant growth. As early as 1915 Mazé had included chlorine in nutrient cultural solutions. F. M. Easton had reported in 1942 that in his tests chlorine had increased the growth of cotton and tomato plants[2]. However, it was the work of the Broyer team that, by applying the rigid criteria of Arnon, proved that chlorine qualified as an essential nutrient element. The same Broyer group of scientists reported in 1957 that they had been able to induce acute chlorine deficiency in lettuce, cabbage, barley, alfalfa, and field beans, besides tomato, sugar beet, buckwheat, and corn. Of all the plant species examined, it seemed lettuce was most sensitive and squash least sensitive to a lack of chlorine.

Broyer and associates had used the tomato plant in their investigation. Subsequently, others used sugar beet, barley, and alfalfa as test

[1] Broyer, T. C., Carlton, A. B., Johnson, C. M., and Stout, P. R., *Plant Physiol.*, **29**, 526–532 (1954).
[2] Eaton, F. M., *J. Agr. Res.*, **64**, 357–399 (1942).

plants. Ozanne, another investigator, using two soils from California and one from Australia, which analyzed 1–2 ppm Cl, induced chlorine deficiency in subterranean clover grown on these soils under greenhouse conditions[3].

Establishing the essentiality of a nutrient element is no simple matter. One of the principal problems is to prevent contamination of the nutrient solution. Contamination occurs from many sources: from the culture media, from aeration, from the specific nutrient naturally present in the seeds of the test plant, and from other uncontrollable sources. These difficulties were also operative in the case of the micronutrient chlorine. The difference, however, was that chlorine was present in plant tissues in amounts measured in parts per hundred, whereas all the others (iron excepted) were needed in parts per million or billion. Hence, it surprised many when it was accepted in the status of a micronutrient element.

Chemistry of Chlorine

Chlorine (Cl) is the most abundant of the halogens. It occurs as the chloride ion in sea water, salt beds, and salt wells, in which situations it is combined with the cations sodium $(Na)^+$, potassium $(K)^+$, magnesium $(Mg)^{2+}$, and calcium $(Ca)^{2+}$. It is a greenish-yellow gas with its name derived from the Greek, *chlorus*, meaning green. Its atomic number is 17 and its atomic weight 35.457.

Commercially, chlorine is obtained by subjecting aqueous or molten sodium chloride to electrolytic oxidation. Industrial use of the element is primarily as a bleach for paper or wood pulp.

Its most important compounds are those which correspond to valences or -1, $+1$, $5+$, and $7+$, although it also exhibits valences of $2+$, $3+$, and $6+$. The -1 state is most common as Cl^- in hydrochloric acid (HCl). The $+1$ state is in hypochlorus acid (HOCl), and its salts, the hypochlorites.

Chlorine occurs in most soils and plants. In plants it exists largely in the form of water-soluble chlorides.

Unlike other nutrient elements present in native rocks, chlorine does not seem to be fixed by soil colloidal materials; in fact, owing to the negative charge carried by chlorine ions (Cl^-), they are repelled from the negatively charged soil colloidal complex. Furthermore, the

[3] Ozanne, P. G., *Nature*, **182**, 1172 (1958).

chlorine compounds formed in soil appear to be highly soluble, which is unlike the reaction of all the other recognized micronutrients.

Anions in the soil solution may interfere with one another during absorption by a plant's roots. A study is reported[4] in which the rate of uptake of chlorine ion by perennial ryegrass is compared when phosphate and sulfate ions are present in solutions of equal ionic concentration. In the test one of the anions was gradually replaced by another. The effect was reflected in the anionic composition of the ryegrass; as the rate of uptake of the chlorine ion increased, the uptake of PO_4^{3-} and SO_4^{2-} ions decreased in that order.

In a pot experiment Dijkshoorn[4] found that ammonium nitrate depressed the Cl content of perennial ryegrass, especially when a chloride was also added. Another investigator, Rahman, applied ammonium nitrate at the rate of 336 lb N per year, with the result that the Cl content in the crop was reduced by about 25%, the reduction seeming to increase when the potassium ion (K^+) was present. The chlorine contents in these Rahman tests ranged from 0.38–0.63%, dry weight. Rahman also observed that the K^+ ion, even when added as K_2SO_4, tended to increase the Cl content, although this effect diminished as the sward became established[5].

Sources of Chlorine

In farming areas the common sources of chloride are fertilizers, particularly potassium chloride and ammonium chloride, sewage, animal manures, and plant residues. In some parts of the world salt spray from seawater that has been carried inland by winds to a depth of about four miles is a constant source, although not always beneficial, to commercial crops.

Many authorities doubt whether chlorine deficiency will ever become a significant factor in the growth of most commercial crops because of its widespread prevalence and numerous sources of supply.

Chlorine in Plants

Fleming has published data[6] which indicate that certain distribution patterns exist in grasses, but the degree to which they occur varies

[4] Dijkshoorn, W., *Neth. J. Agr. Sci.*, **6**, 131–135 (1958).
[5] Rahman, H., et al., *J. Sci. Food Agr.*, **11**, 422–428 (1960).
[6] Fleming, G. A., *Outlook Agr.*, **6**, 283 (1965).

widely among species; the level of Cl content is always higher in stems than in the leaves or heads. For example, the following data serve to illustrate. The grass species are rough stalked meadow grass (Poa trivialis) designated by A; crested dogstail (Cynosurus cristatus) by B; and Italian ryegrass by C:

Grass	Plant Part	Chlorine Level %		
		A	B	C
A, B, C	Head	0.28	0.24	0.31
	Leaves	0.71	0.45	0.43
	Stem	0.75	0.75	0.48

From many studies it is concluded that, as in the case of phosphate, a large number of plant species seems to assimilate Cl at maximal rate during the early stages of the growth cycle.

Berger has reported that potato leaves in Wisconsin will contain, on the average, as much as 5.5% Cl, dry weight[7].

Fleming has also reported[6] on the Cl content of six grass species grown separately on plots that received potassium chloride fertilizer at the rate of 200 cwt KCl/acre, in addition to a basal dressing. The results were as follows:

Yorkshire fog (common velvet grass) 1.0%
crested dogstail 0.71%
rough-stalked meadow grass 0.45%
bent grass 0.45%
Italian ryegrass 0.35%
red fescue 0.32%

Potassium chloride (KCl) fertilizer seems to influence greatly the Cl contents of pasture herbage. Kemp[8] applied 340 lb K/acre, as 60% KCl, and found an increase in the seasonal average K content of grass herbage from 1.28–1.86%. But when he applied 100 lb P/acre, as ammonium phosphate, the effect of the Cl content of the herbage was insignificant.

Analyses of pasture herbage shows a wide variation in Cl content. Orr reported[9] results from the analyses of 200 samples ranging in Cl content from 0.015–1.71% Cl, dry weight. In 101 samples of Italian

[7] Berger, K. C., "Introductory Soils," New York, Macmillan Co., 1965.
[8] Kemp, A., Neth. J. Agr. Sci., 8, 281–304 (1960).
[9] Orr, J. B., "Minerals in Pastures," H. K. Lewis and Co., 1929.

ryegrass taken from 13 different locations in Great Britain, Orr found that the Cl content ranged from 0.41–1.99%.

Although it is not often that plants are found suffering from a deficiency of chlorine, it does happen in some areas, as for example, on sandy soils far from the sea in California and Australia.

Excessive amounts of chlorine may be absorbed, as occurs in Puerto Rico (to name one of many areas). Here the Northeast trade winds carry sufficient sodium chloride in the form of a fine spray to areas about four miles in from the coast to cause a heavy intake of Cl by the tobacco crop; while this does not harm the tobacco plant itself, it does have a damaging effect on its quality, for an excessive amount of chlorine in the tobacco leaf harmfully affects its burning quality.

Stout has shown that tomato plants suffering from a severe lack of chlorine will have, in their leaves, an average of about 250 ppm Cl, dry weight. Molybdenum deficient plants showing equal visual deficiency symptoms contain an average of about 0.1 ppm Mo, dry weight. Hence, he calculates that on a weight basis the amount of chlorine is several thousand times greater than the amount of molybdenum. On an atom-for-atom basis, since 3 chlorine atoms weigh as much as one molybdenum atom, the chlorine requirement is about 8,000 times greater than that of molybdenum[10].

It is estimated that plants generally require about 1 lb of chlorine for each 4,000 lb dry matter they produce. Large crops, accordingly, would need approximately a minimum of 5 lb Cl/acre.

The amount of chlorine commonly found in soils in the form of soluble chlorides will vary, and estimates range from 100–1,000 lb Cl/acre. Berger states that in Wisconsin soils the average amount of chlorine ranges from 0–74 lb/acre, with an average of 34 lb/acre in the plow layer and small amounts in the subsoil[11]. Soils of arid regions generally contain more chlorine because less is leached out. Saline soils have, as expected, a relatively high Cl content, and the chlorine will be responsible in most cases where damage is suffered by crop plants.

Deficiency Symptoms

Plants suffering from a severe lack of nutrient chlorine exhibit wilting of leaflet tips, a progressive chlorosis of leaves, followed by bronzing,

[10] Stout, P. R., *U.S. Dept. Agr. Yearbook*, 147 (1957).
[11] Berger, K. C., ibidem.

12.1   Chlorine deficiency symptom in tomato leaves. Left: leaf severely deficient. Center: Leaf less deficient in chlorine due to partial replacement of chlorine by a bromide. Right: Growth favorable; supply of chlorine adequate. (*Courtesy of T. C. Broyer, University of California*)

and finally the leaves die (see Figure 12.1). In severe cases the plants fail to form fruit. In the sugar beet an interveinal chlorosis first appears in the middle leaves, somewhat similar to the effect of manganese deficiency. The leaves take on a mottled effect which becomes visible only by transmitted light; these affected areas later appear as light green to smooth and flat depressions. In some cases the secondary roots become stumpy.

Chlorine in Animal Nutrition

Cl is essential in animal nutrition, for which purpose the element is utilized chiefly in the form of sodium chloride or common salt. Both ions, sodium (Na^+) and chlorine (Cl^-), are closely associated in their biological roles and are particularly important in the maintenance of proper osmotic concentrations in fluids and cells[12]. Cl is also involved indirectly in the transport of CO_2 by the blood, and as an ion, it serves as a major electrolyte in controlling the solubility of proteins, especially the globulins. About two thirds of the anions contained in the bloodstream are chlorine ions, and a normal Cl concentration in the serum is about 0.36%, while in red cells it ranges from 0.19–0.23%. Urine contains an average of about 0.6% Cl ions, the amount depending upon the total amount ingested and the volume of urine.

Normally humans and animals require Cl in the diet, but the precise amount needed is difficult to ascertain. Since herbage plants tend to be somewhat low in Cl^- and Na^+ content, it is necessary to supplement salt in the diet of herbivores. On the other hand, animal tissues are high in the content of these two ions, and carnivores do not require supplementary salt.

De Groot[13] suggested that, especially when the herbage is high in K^+ content, the animal metabolism would be improved by feeding herbage with a correspondingly high Cl^- content. He considers levels of less than 0.5% Cl as undesirable for feed[13]. Animal requirements of Cl are satisfied by a content of about 0.19% in the diet.

Unless salt is made available, it is quite possible for chlorine deficiency to develop in cattle, sheep, and even deer. Even horses may need supplementary salt in hot weather if they sweat profusely while working. Humans are also inclined to develop a deficiency under similar circumstances; the loss of sodium chloride may be at a rate of 2 gm/hr, and this loss must be restored if the perspiration continues for some time. Another serious loss may occur in cases of diarrhea or vomiting.

Symptoms of Cl deficiency in the animal system are severe abdominal cramps, weakness, rapid loss of weight.

Normally, of the total ingested Cl in the animal, about 90% is excreted in urine, 5% in perspiration, and about 1% in the feces. These figures will vary since they will be influenced by the rate and amount ingested, the muscular activity, the liquid intake, and the temperature of the environment.

---

[12] Mallette, M. F., "Biochemistry of Plants and Animals," New York, John Wiley and Sons, 1960.
[13] Groot, T. de, *Stikstof*, **8**, 33–40 (1964).

Fleming summarizes his conclusions as follows:

"The trace elements in pasture species must satisfy both the nutritional requirements of the plant and the needs of the grazing animal. The growing plant may suffer from deficiency or excess of any given element; absorption of one metal ion may disturb the entry of another; species differences in uptake and tolerances exist, and may vary with the season. On the balance of these factors, the health of the animal depends"[14].

[14] Fleming, G. A., ibidem (1965).

13

SODIUM

Sodium (Na, the symbol for Natron) is one of the alkali metals group of chemical elements, which includes lithium, potassium, and rubidium. It occurs in nature only, with a valence of $+1$ (Na^+). Its atomic number is 11, and its atomic weight is 22.991. Sodium is sixth in abundance among the earth's mineral elements, but despite its abundance it occurs in nature only in combined form, as in common salt ($NaCl$), Chilean nitrate of soda, borax, albite, and diorite. Sodium is readily soluble in water and is found in seawater and in brine wells. When it reacts with oxygen, it forms the peroxide (Na_2O_2), and with water, the hydroxide, sodium hydroxide ($NaOH$) or caustic soda. Other important compounds are washing soda (Na_2CO_3), and baking soda ($NaHCO_3$).

The sodium ion is an indispensable constituent of plant and animal tissues, being the principal cation of the fluids *outside* the cells as potassium is of the cations *inside* the cells. The sodium ion is required for contraction of all animal muscle. As plant food, sodium improves plant vigor, helps resist diseases, adds flavor to many foods, and improves the keeping quality of many crops.

Recent published data have shed much light on the role of sodium as a plant nutrient. In the early days of agricultural chemistry, sodium was regarded as an element essential to plant life. About 1860 this view fell from favor, and sodium was considered an inert carrier or a stimulant capable of acting as a partial substitute for potassium. Between 1860 and 1935 its role was quite uncertain; some claimed it merely served to liberate potassium from its soil combinations, or that it released potassium in various parts of the plant, making it available for use elsewhere, or that it fulfilled a direct, essential role in plant nutrition. From 1935–1950 the physiological role was accepted, but while it was considered a plant nutrient, it was not thought of as being essential. About 1950 most investigators conceded there are functions in plant life for which sodium is in fact essential.

It is now accepted that sodium is essential for the blue-green alga, *Anabaena cylindrica*, and for *Atriplex vesicaria*, one of the *Chenopodiaceae*.

The sodium ion (Na^+) is closely associated with the chlorine ion (Cl^-) in most of its activities in plant and animal metabolisms. The need for the element sodium in animal life is the key to salt requirements. In humans sodium helps maintain the vital neutrality of the circulating blood, regulates water balance so that the body becomes neither waterlogged nor desiccated. It is estimated that to protect the body against acidity, over 90% of the basic ions in the bloodstream are sodium.

As for the apparently beneficial results obtained when a sodium salt has been applied in the fertilization of several crops, numerous explanations have been proposed by researchers. As previously mentioned, sodium has been known for many years as an element of great value to a wide variety of crops and to the quality of animal feedstuffs produced on the farm. However, it has been more or less neglected as a valuable plant food, as is apparent to the investigator who studies the literature.

List of Benefits[1]

With the above in mind it is interesting to record the following list of benefits attributable to sodium:

(1) Sodium can help to replace potassium in many crops, including sugar beet, red beet, oats, cabbage, kale, turnips, mangels, carrots, Swiss chard, cereal crops, and cotton. In the beet family it can

[1] Ferguson, J. K., *Fertilizer Feeding Stuffs J.*, **29**, 323 (1964).

replace up to half of the potassium requirement. Since sodium costs less than potassium, farmers should give it consideration.

(2) Sodium is for many crops an indispensable food and will help to increase yields even when potassium has been applied generously.

(3) During periods of drought it will retard wilting because of its capacity to attract moisture from morning dew, the atmosphere, and from the lowered water table.

(4) During the winter and early spring, sodium may help control frost damage by lowering the freezing point of plant moisture. This helps over-wintered crops and crops that appear in spring when frost is still a possibility.

(5) As a metallic cation sodium can reduce the loss of lime from the soil and help stabilize the ionic balance in the soil solution.

(6) It has the ability to disintegrate insoluble phosphate in the soil and render it available to crops; this is especially true on calcareous soils.

(7) It imparts color and flavor to market garden crops such as cabbage and other kohl crops; cabbages retain a bright green color.

(8) Recent research into the effects of sodium applications to pasture herbage and into the subsequent effect on animals grazing it or fed on fodder from the treated swards indicates that the sodium has considerable value for the livestock. Dairy cattle require a daily intake of from 4–5 gm of sodium (Na) per gallon of milk produced, above that needed for maintenance, and 2 more grams for optimal production. These amounts are in addition to the 17 gm required daily for normal bodily needs. A drop in this total daily intake of sodium in their feed will cause a corresponding drop in milk production; and even worse, a deficiency in sodium, it is now believed, may lead to grass tetany.

Sodium and Soil Structure

Crop responses to sodium as sodium chloride fertilizer are better with good drainage than with poor and in a wet season than a droughty one. One of the strong indictments against the use of sodium salts for fertilizer purposes has been that abnormally large amounts of sodium damage the soil structure, making the soil sticky and hard to work. It is conceded that an excessive quantity of sodium may have this bad effect, but sensible rates of application do not and therefore should be adhered to. Experience in most countries indicates that generally the most acceptable form of sodium to apply to most crops is sodium nitrate ($NaNO_3$). A normal dressing of sodium nitrate supplies an amount of sodium that has no appreciable effect on any soil. The application rate for different crops varies but it is agreed that rates of 200–400 lb per acre per year will supply sufficient sodium for most purposes. For example, 200 kg $NaNO_3$ per hectare (176 lb/acre) would add only about 10 mg Na/kg of top soil, about one part in 100,000 parts by weight. The presence of lime ($CaCO_3$) in the soil receiving sodium will strongly tend to offset the action of sodium on the soil's structure.

Sodium in Plants

Soluble compounds of sodium are readily available to plants. The leaves are usually richer in the element than the seeds, and leguminous plants tend to be richer in sodium than grasses. There is evidence that, if plants require it, it is at a tissue concentration of less than 2.3 ppm[2].

Reported Na contents of pasture herbage vary more than 1,000-fold, from 0.002–2.12%, and any values between 0.05–1% would not be considered unusual.

The factors affecting the content of sodium in grass have been recently reviewed by C. H. Henkens[3]. He summarized evidence to the effect that the contents of sodium are governed largely by the Na content and the K status of the soil, the Na content being of greater influence when the K status is low. He and Oostendorp and Harmsen[4] have demonstrated that good correlations exist between the Na content of grass herbage and the K:Na ratio (expressed as the $K_2O:Na_2O$ ratio soluble in 0.1 HCl) in the soil. Oostendorp and Harmsen believe that a soil ratio of 4 or less was necessary if the herbage Na was to be maintained at more than 0.2%.

Phosphorus (P) does not have much effect on the Na content of grasses, although Stewart and Holmes[5] found that Na was generally decreased by the presence of P if nitrogen were absent and increased by P in the presence of N.

The Na content of many grass species is greatly increased by applications of nitrogen, at least when it is added in the absence of

[2] Gerloff, G. C., *Ann. Rev. Plant Physiol.*, **14**, 107–14 (1963).
[3] Henkens, C. H., *Neth. J. Agr. Sci.*, **13**, 21–47 (1965).
[4] Oostendorp, D., and Harmsen, H. E., cited in *Herbage Abstr.*, **34**, 1713 (1964).
[5] Stewart, A. B., and Holmes, W., *J. Sci. Food Agr.*, **4**, 401–408 (1953).

potassium. Reith et al.[6] reported that the mean Na contents of mixed herbage from 4 locations increased from 0.39% with no N to 0.63% with 174 lb N/year and to 0.69% with 348 lb N/year per acre, applied without K.

Sodium in Animals

Sodium is essential to animal life. Hogan and Nierman[7] found the body of the steer contains 0.16% Na. The body of a 225 lb hog contains 0.16 lb Na equivalent to 0.07% Na. Sodium is present almost entirely in the soft tissue and fluids of the body; in the blood, it forms 93% of the bases of the blood serum. Sodium has to do with maintaining the acid–base balance within the body, with the activity of the muscles, with osmosis and the retention of water by the tissues.

Growth requirements for rats, chicks, pigs, and calves amount to about 0.1–0.2% of the ration, the exact amount depending on the species and other factors of the diet[8].

Excessive amounts of Na are excreted as chloride by the kidneys. The general use of common salt as a condiment makes deficiencies of sodium rare in humans except in certain diseases or in prolonged, profuse sweating.

Sodium as Fertilizer

An important study supporting the argument that more attention should be paid to the responses of crops to sodium was published by Dr. W. P. Mortensen, formerly of the University of Wisconsin[9]. In it he reviewed the results of hundreds of plot, greenhouse, and field tests on a wide variety of soil types; the number of crops which responded to salt applications exceeded substantially the number that did not, 22 against 9. Sugar beet was an outstanding crop example. Dr. Mortensen got significant yield increases from 300 lb of salt/acre for wheat, barley and oats; these were, respectively, an extra 303, 6.2 and

[6] Reith, J. W. S., et al., *J. Agr. Sci.*, Cambridge, **63**, 209–219 (1964).
[7] Hogan, A. G. and Nierman, J. L., *Mo. Agr. Etx. Stn. Res. Bull.*, 107 (1927).
[8] Mallette, F., et al., "Biochemistry of Plants and Animals," New York, John Wiley and Sons, 1960.
[9] Anon, *World Crops* (June, 1955).

2.4 bu/acre. Celery also showed significant response, and the use of 300–800 lb salt/acre is now a standard recommendation in many celery growing areas to improve quality and yields.

James Dundas presented a strong case for the profitable use of sodium in raising sugar beet in Britain[10]. He showed that sodium is more important for sugar beet than potassium. In continental Europe the value of sodium is appreciated. Nine out of every ten Swedish sugar beet growers use it in the form of sodium nitrate. Adams[11] concluded from the results of many tests that although potassium is a nutrient for sugar beet, its chief value is, on most British soils, as a substitute for sodium. Russell[12], discussing sugar beet experiments at Rothamsted Experiment Station, said: "Thus there is a definite need for sodium, and given the sodium there is no need for potassium; but in its absence potassium can in part take its place."

Farmers on muck soils in Michigan use sodium very extensively as fertilizer on celery, table beets, and sugar beets. Most of the big growers of table beets for canning in the Geneva, New York area use sodium regularly as a fertilizer. The sodium improves the quality and flavor of the beets. The color and flavor of the beet tops are also improved, which is of interest to those who use them for greens. On mineral soils at the Geneva Experiment Station, sodium increased yields of table beets even when 200 lb of K_2O/acre was provided.

[10] World Crops, 220–224 (July, 1962).
[11] Adams, S. N., Brit. Sugar Beet Rev., **27**, 111–113 (1959).
[12] Russell, E. J., "Soil Conditions and Plant Growth," Longmans, Green and Co., 44, 1961.

14

SELENIUM

Selenium (Se) has been recognized since 1957 as a nutrient element essential to animal life. Some slight evidence that it may also be essential to plant life has been accumulated, at least it is known to be beneficial to certain genera. For many years selenium was considered a poisonous element fatal to livestock on the western ranges.

It is believed that the first authentic record of selenium poisoning in livestock was made by T. C. Madison in 1856 in a report to Congress[1]. He described a fatal disease in army horses in Nebraska characterized by loss of hair from the mane and tail and soreness of the feet. A similar condition was later described as occurring in cattle and sheep and given the name "alkali disease," because it was attributed to drinking an excessive amount of alkaline waters. Another form, but more acute, of the condition was named "blind staggers." These names have persisted and others coined, although the theory of cause has long been disproved. The modern view of this condition attributes it to selenium poisoning, and it can affect cattle, sheep, horses, swine, and poultry. The symptoms of the disease include loss of hair in horse and cattle and of hair from the body of pigs, and changes in the growth of the hoof in

[1] Madison, T. C., *Senate Executive Doc.*, Jan. 1855–Jan. 1860; **52**, 37–41 (1860).

186

all these animals, which, in severe cases, can result in the sloughing of the hoof. The disease is also known now by the name, "white muscular disease" (WMD) and "muscular dystrophy," and additional symptoms have been recognized, including atrophy of the heart, emaciation, anemia, excessive salivation, grinding the teeth, and death caused by inability to breathe. Affected poultry eggs do not hatch. Up till now no case of selenium poisoning has been identified in humans.

Thus selenium, recognized for about a century as toxic to domestic animals, was in 1957 demonstrated by Schwarz and Folz[2] to be a trace element essential to animal life.

Certain soils of the Great Plains Region derived mainly from rocks of the Cretaceous Age were found to contain selenium in less than 10 ppm Se, but soils considered toxic in the USA and Canada contain between 1 and 6 ppm Se in the top 8-in. layer. Much higher levels have been determined in some soils. Soils having a high toxic content of selenium are often of tuff, shale, or limestone groups. A profile of rock frequently shows a wide variation of selenium content with depth. Considerable quantities of selenium are carried off by the drainage waters of certain western streams. Rainfall of more than 25 in./year leaches out the readily soluble and available Se. Irrigation waters will also, in time, leach sizeable quantities of selenium from the soils.

For a long time scientists suspected that certain plants absorbed selenium, but it was not until 1933 that it was positively shown they do so. In that year quantitative estimates were made in the U.S. Department of Agriculture[3]. The literature now contains reports of thousands of a wide variety of both native and crop plants. Of the samples reported by Robinson in 1933, two contained levels of Se toxic to rats, namely, 5 and 11 ppm. From many published analyses of a number of farm crops grown in the seleniferous soils of Western United States, the general levels of Se in the majority of samples collected range from 0.1–4 ppm. In the course of these studies it became evident that typical selenium accumulator plants were commonly found on seleniferous soils, such as species of the genera *Astragulus* (a legume) and *Stanleya* (a mustard), *Xylorrhiza* (woody aster) and *Oonopsis* (related to golden rod).

The form of selenium in plants has not, as yet, been established, although some reports indicate that it is present as a seleno–amino acid similar to cystine and leucine.

[2] Schwarz, K., and Folz, C. M., *J. Am. Chem. Soc.*, **79**, 3292 (1957).
[3] Robinson, W. O., *Assoc. Off. Agr. Chem. J.*, **16**, 423–424 (1933).

Chemistry of Selenium

Selenium is a "metal" whose chemical relationships are similar to those of sulfur. Its atomic number is 34 and atomic weight, 78.96. It reacts with metals to form selenides, for example, Al_2S_3, which decompose in acid to give hydrogen selenide (H_2Se), which, like H_2S, is toxic and burns to give Se or SeO_2. H_2Se is a stronger reducing agent than H_2S. SeO_2 is a colorless solid which dissolves in water to give the weak acid, selenious acid (H_2SeO_3). This acid can be oxidized to give selenic acid (H_2SeO_4), the analog of sulfuric acid (H_2SO_4).

Research and "Factor 3"

The research which ended with the discovery of Se as an essential nutrient factor is quite interesting. The project originated in studies of brewer's yeast as a protein supplement in Europe during World War II. German scientists found that rats on a yeast diet developed a liver degeneration, and that wheat germ and wheat bran protected against this disorder. It was subsequently proved that in fact it was vitamin E that was the main protective agent.

American researchers at the same time reproduced the German results by using a diet based on torula yeast[4]. Schwarz named the unidentified effective material present in American yeast, "Factor 3," but it is not present in torula yeast or in European brewer's yeast.

Later Scott and associates[5] showed that brewer's yeast prevented exudative diathesis in poultry, a disease recognized as being caused by vitamin E deficiency, and which could be induced by a torula yeast diet[5]. Selenium was finally, in 1957, identified as the key component in Factor 3, although the precise mechanism of its metabolic role with vitamin E has as yet not been clarified. McLean, one of the first to recognize growth responses to selenium, summarized the current status as follows:[6]

> "There can be little doubt that traces of selenium are required by the animal for normal metabolism, that vitamin E and selenium are interrelated in the metabolic functions, and that Vitamin E cannot completely replace the need for selenium."

[4] Schwarz, K., *Proc. Soc. Exp. Biol. Med.*, **78**, 852 (1951).
[5] Scott, M. L., *Proc. Cornell Nutr. Conf.*, 111 (1958).
[6] McLean, J. W., et al., *New Zealand Vet. J.*, **7**, 47 (1959).

Se Supplementation

Important reports on selenium supplementation have been published by New Zealand investigators, a country where extensive land areas are deficient in Se. By means of small doses of selenium preparations several disorders have been corrected, such as white muscle disease in cattle, sheep (stiff lamb disease), horses, and swine, and exudative diathesis of poultry, and "ill thrift," infertility, and chronic scouring of sheep.

It is now known that as little as 0.05–0.1 ppm Se can prevent muscular dystrophy in lambs and calves.

In 1934 Beath and associates[7] reported that some poisonous Wyoming range plants, including *Astragulus bisulcatus*, were capable of storing 1,000 ppm Se when grown on certain cretaceous derived soils. This removed any doubt in the ability of some plants to accumulate toxic amounts of Se from natural soils and also provided an explanation of why heavy losses of cattle and sheep occurred on the ranges of the Great Plains (see Figures 14.1, 14.2, 14.3). According to Sullivan and Garber[8], the tolerance limit for cattle and sheep is 4 ppm Se.

Rosenfeld and Beath[9] indicate that Se poisoning can be caused by grass and hay containing 10–30 ppm Se. At Rukuhia Soil Research Station (New Zealand), it was found that pasture herbage generally contains a low content of Se, for example, brown top> ryegrass> cocksfoot> red clover> white clover, and in general the Se content was of the order of 0.01–0.1 ppm. The following table illustrates this variation in uptake.

Apparently, the relative uptake is influenced by the Se supply. Cultivated crops and native grasses seldom contain enough Se to cause harm to animal life. The primary problem, as previously mentioned, is posed by Se accumulating plants, among which species of the genus *Astragulus* are the most notorious, some being able to store up to 15,000 ppm Se. Species of this genus are widely distributed, 24 of its more than 200 species are known to be Se accumulators, and these seem to need it in their diet to grow normally[10].

---

[7] Beath, O. A., et al., *J. Am. Pharm. Assoc.*, **23**, 94–97 (1934).

[8] Sullivan, J. T., and Garber, R. J., Pa. Agr. Expt. Sta. Bull., 489 (1947).

[9] Rosenfeld, I., and Beath, O. A., "Selenium," New York, Academic Press Inc., 1964.

[10] Bear, F. E., *U.S. Dept. Agr. Yearbook* (1957).

TABLE 14.1   DIFFERENTIAL UPTAKE OF Se BY GRASSES AND CLOVER[11]

TOP-DRESSING RATE OUNCES Se/acre	SELENIUM CONTENT ppm			
	BROWNTOP	RYEGRASS	COCKSFOOT	WHITE CLOVER
0	0.022	0.012	0.012	0.008
2	0.28	0.23	0.16	0.08
4	0.45	0.25	0.18	0.08
8	0.84	0.46	0.36	0.21
16	1.54	0.86	0.68	0.34

Treatment

As would be expected with an element which has been described as "the only mineral element known to be absorbed by food and forage plants in sufficient amounts to make them lethal when consumed by animals," the amount of selenium required for beneficial treatment is minute. Under controlled experimental conditions in New Zealand, five different animal disorders, all of economic importance, have been prevented by treatment with Se[12]. The disorders are as follows: white muscle disease (WMD) in lambs, barrenness of ewes, ill thrift in lambs and calves, and exudative diathesis in chicks.

The doses required are too small to permit incorporation with feedstuff as supplements, so individual dosing of the animals now seems the only safe method. The New Zealand recommendations are as follows: to prevent both the infertility and WMD problems in sheep, the dose is 5 mg of Se by mouth one month before mating and repeated one month before lambing; for ill thrift in lambs the dose is 1 mg at tailing and 5 mg doses once every three months; for dairy calves, give twice this amount once every three months; for poultry, give 10–15 mg per gallon of drinking water one day per month.

The report emphasizes that these Se requirements apply only to farm animals raised on soils having selenium deficiency.

The Department of Agriculture and Veterinary Club officers of New Zealand, where WMD and ill thrift have prevailed for some years,

[11] Hartley, W. J., Grant, A. B., and Drake, C., *New Zealand J. Agr.*, **101**, 343 (1960).
[12] Hartley, W. J., Grant, A. B., and Drake, C., *New Zealand J. Agr.*, **101**, 343 (1960).

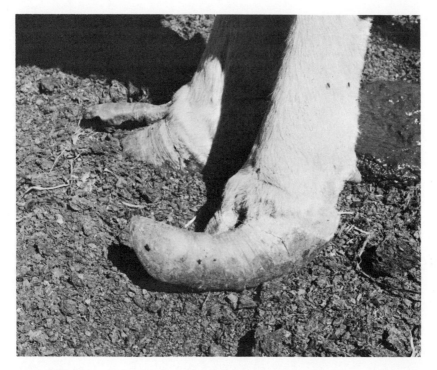

14.1   Selenium toxicity, showing effect on animal's hoof. (*Courtesy of So. Dakota Agr. Expt. Station*)

have proved by extensive trials on both islands that controlled doses of selenium can correct these two disorders. Along with other toxic mineral micronutrients such as cobalt, molybdenum, chromium, vanadium, and arsenic, the use of selenium as a prophylactic and curative agent presents a real problem when used for livestock purposes. Under the strict veterinary control the likelihood of trouble is much lessened. A bad feature of selenium is its delayed action in the poisoning of livestock. Experience in the United States and elsewhere indicates that under certain not fully understood conditions, an animal may not show any outward sign of poisoning, perhaps for several months, after grazing on seleniferous range plants. For example, it is known that cattle and sheep, which seem to be sound and healthy but may suddenly go "off feed," pass bloody urine, and rapidly lose weight, may have grazed the toxicant herbage some months previously. Since selenium is more

14.2 Selenium toxicity symptoms in cattle. (*Courtesy of So. Dakota Agr. Expt. Station*)

toxic to humans and animals than to plants and is not easily detected without the use of indicator plants and special analytical procedures, it is of great importance in its relation to public health.

The safe limits of selenium by humans and animals have not as yet been completely defined. Although selenium compounds have been proposed and some are being used quite successfully for plant, pest, insect, and nematode control, the gap between safe and toxic limits is very narrow and suggests that they be employed with great care.

Results of successful feed supplementation with selenium in New Zealand led to a survey of the status of selenium research at the United States agricultural experiment stations[13]. This has shown that selenium deficiency diseases such as white muscle disease of cattle and sheep are widespread, although often unrecognized. The correction of subacute selenium deficiencies has given remarkable gains in weight, fertility, and survival rates. In Oregon, ewes' rations were given a 0.1 ppm Se supplement in the form of sodium selenite ($Na_2SeO_3 \cdot 5H_2O$), and this

[13] Wolf, E., et al., *Agric. Food Chem.*, **4**, 355–360 (1963).

14.3   Selenium toxicity symptoms in sheep. (*Courtesy of U.S. Dept. Agriculture*)

consistently prevented WMD, whereas vitamin E supplementation did not. Generally, in those areas of the United States where Se is being administered to cattle, the principal method is by injection and oral drenching. Selenium treatment should be given only under specialized direction. Other Se compounds are being investigated, such as barium selenate and an organic compound of selenium. Direct soil application may eventually become the most effective, lowest cost method of correcting Se deficiencies in livestock; a great deal more research will have to be done on this method before it can become a recommended practice.

Investigators proved that it was possible to produce the known manifestations of Se poisoning by experimentally feeding selenites.

Selenium was shown to be capable of replacing sulfur, which it closely resembles in chemical reactions, in the amino acids cystine, gluta-thione, and probably methionine[14].

In 1957 Schwarz and Folz[15] reported the presence of organically bound Se in preparations of what Schwarz had previously named "Factor 3." Inorganic Se was proved to protect rats completely from the development of necrotic liver degeneration, which Weichelsbaum had previously recognized as a Se deficiency disease[16]. From this research Schwarz and Folz stated:

> "It can be inferred from our results that selenium is an essential trace element." They also suggested, on the basis of other experimental work, that selenium may function in oxidation-reduction reactions.

In 1957 two groups of workers[17] reported that Se added in minute amounts to diets known to produce exudative diathesis in chickens gave complete protection.

These reports and similar following reports by other scientists resulted in the acceptance of selenium by the scientific world as an essential trace element for animal life. According to conclusions inferred by Beeson from a mass of experimental data, the evidence so far available is that selenium is not an essential element for plant growth, even for the selenium indicator plants such as the species of *Astragulus*[18].

Selenium as Insecticide

Gnadinger[19] first investigated selenium compounds as possible insecticides in 1933, one reason being its close relationship to sulfur, at the time used as a fungicide and miticide. The Se compound selected for use as a miticide on greenhouse plants was made by dissolving selenium in potassium ammonium sulfide solution in proportions corresponding to the formula $(KNH_4S)_5Se$. A 30% solution designated Selocide used at a 1 to 800 dilution with a soap spreader gave good control of red

[14] Todd, W. R., *J. Agr. Food Chem.*, **3**, 166–173 (1959).
[15] Schwarz, K., and Folz, C. M., *J. Am. Chem. Soc.*, **79**, 3292 (1957). Schwarz, K., *Proc. Soc. Exp. Biol. Med.*, **78**, 852 (1951).
[16] Weichelsbaum, T. E., *Quart. J. Exptl. Physiol.*, **25**, 363.
[17] Scott, M. L., et al., *Poultry Sci.*, **36**, 1155; and Stockstad, E. L. R., et al., *Poultry Sci.*, **36**, 1060 (1957).
[18] Beeson, K. C., "Selenium," in Agric. Handbook, 200, USDA & USDI, 1961.
[19] Gnadinger, C. B., *Ind. Eng. Chem.*, **25**, 633–637 (1933).

mites on citrus. Residues of selenium in fruits of treated grapes and citrus, however, tended to discourage its use.

Sodium selenate as crystals usually dyed blue or red is available in bulk to florists and nurserymen who make up solutions for soil drenches to control spider mites, aphids, and certain nematodes.

Capsules containing $\frac{1}{4}$ gm of sodium selenate are packaged for shipping by mail or for sale in seed stores and garden supply houses. They are used mostly in the home for control of cyclamen mites on African violets or other plants.

The total quantity of selenium insecticide compounds actually marketed annually is small. California seems to be the only state in which seleniferous insecticides have been registered for use on food crops.

Health Hazards

In 1933 Nelson and others[20] cautioned growers against the use of Se as an insecticide. As little as 1 ppm in the soil permitted the growth of wheat, but when the grain containing 8–10 ppm Se was fed to rats, it retarded growth and killed them after a few weeks.

In New Zealand Taylor[21] pointed out that the maximal amount of Se in the whole diet that man can consume without ill effects is believed to be 3 ppm. A daily intake of 1 mg Se in food is probably not harmful to an adult person.

Fuller[22] emphasized that selenium or its compounds should in no case be used for food for humans or domestic animals or on land that might be used for growing such crops. Present official U.S. recommendations for use of selenium stipulate it has a zero tolerance on all food crops and has been registered for use only on ornamentals, except for the registered use on citrus in California.

Analytical Methods

Because animals require such very minute amounts of selenium to correct deficiencies, it is necessary to employ extremely accurate

[20] Nelson, E. M., et al., *Science*, **78**, 124 (1933).
[21] Taylor, G. G., *New Zealand J. Sci. Technol.*, sec A 34, **34**, 36–46 (1952).
[22] Fuller, Glen, *Chemurgic Digest*, **6**, 113 (April 15, 1947).

methods of analysis. In developing such analytical procedures good use was made of the available radioactive sources and sensitive procedures of neutron activation analysis. This method has proved very useful for determining the Se content of plant and animal tissues. Unfortunately only a few laboratories have the required special equipment and trained personnel. The practical limit of sensitivity is said to be 0.01 microgram of selenium[23].

It is expected that more use will be made of a fluorometric technique, using 3, 3–diaminobenzidine and sensitive to 0.02 microgram Se[24]. The technique has been improved[25].

[23] Anderson, M. S., *U.S. Dept. Agr. Handbook*, **200**, 24–26 (1961).
[24] Watkinson, J. H., *Anal. Chem.*, **32**, 981–983 (1960).
[25] Hartley, W. J., and Grant, A. B., *Federation Proc.*, **20**, 679 (1961).

15

COBALT

Cobalt (Co) is a trace element whose place in agricultural chemistry has been somewhat anomalous for a long time. Many animal diseases, known by various local descriptive names, are now known to be caused by cobalt deficiency. Originally crude iron salts were employed to remedy the trouble, and it was not realized, until careful Australian research proved it, that the real beneficial factor was the cobalt present in the salts as an impurity. Since then, the application of cobalt sulfate to pastures is the established practice when any of the cobalt deficiency diseases occur. That cobalt is an essential nutrient element for plant life has not as yet been established, although it is found in trace amounts in all foliage. It appears strange that an element as essential to animal life should not also be essential to the life of plants that are natural carriers of it.

Cobalt is required by *Rhizobium*, the symbiotic bacterium that fixes nitrogen in the root nodules of leguminous plants. The requirement is minute, and it is known that many legumes grow normally and fix nitrogen even when the Co content in the herbage is less than 0.08 ppm. The usual range for pasture is 0.03–0.05 ppm.

Cobalt is essential for animals, with nonruminants requiring it in the form of vitamin B_{12}. Ruminants have the advantage of bacterial synthesis of this vitamin from other Co compounds. For ruminants the Co requirement has been estimated at about 0.1 ppm. Apparently in the soil–plant–animal chain, the plant acts primarily as an accumulator of Co for conversion into vitamin B_{12} by organisms, and subsequently the B_{12} then plays an essential role in animal nutrition.

Two relatively recent scientific papers have produced evidence of the importance to plant life, although indirectly, in one[1] for alfalfa, in the other[2] for soya beans (both leguminosae). In both cases these legumes were grown in nutrient solutions without a supply of nitrogen being added, so that the plants utilized only the fixed nitrogen furnished by their nodular bacteria.

Reisenaur showed that furnishing small amounts of cobalt leads to a large increase (over 60%) in weight and a substantial increase in the plant's content of nitrogen. The other researchers give evidence which strongly indicates that cobalt is an essential nutrient for the legume relying on nodular fixed nitrogen for its total supply of nitrogen. Soya bean plants without cobalt added were severely retarded in growth and exhibited severe nitrogen deficiency symptoms; 25% of them died as a result of the cause behind these symptoms. Adding cobalt traces to the solutions of deficient plants cured the symptoms on all new growth within 10–21 days. The trace amounts of cobalt in the culture solutions increased the dry matter in shoots by 23–52%. The actual amount of cobalt used as an addition in the soya bean experiments was of the order of one part in a billion parts of the solution.

A relatively modern fact about cobalt is its association with the molecule of vitamin B_{12}; it is an integral constituent of the molecule. Hence, if cobalt is lacking or the supply inadequate, the production of the vitamin is seriously handicapped. Evans analyzed the bacterial nodules on the soya bean plant roots for their vitamin B_{12} content. Some evidence developed that the nodules from plants receiving cobalt had a higher content of the vitamin. It seems fair to imply that the symbiotic bacteria require cobalt to make vitamin B_{12} for their own needs, and if it is not available they will not thrive and will fix less nitrogen. Other investigators[3] have shown that clovers generally contain more cobalt than grasses. Perhaps it may be stated, on the basis of these recent researches, that cobalt is an essential plant nutrient, at least for leguminosae.

[1] Reisenaur, H. M., *Nature*, **186**, 375 (1960).
[2] Evans, H. J., and Ahmed, S., *Soil Sci.*, **90**, 205 (1960).
[3] Mitchell, R. L., *Soil Sci.*, **60**, 63 (1945).

Chemistry of Cobalt

Cobalt (Co) is one of the elements in the iron triad of the Periodic Table, the other members being nickel and of course iron. Cobalt is much less abundant in the earth's crust than iron and nickel (Cobalt at 0.002% and nickel at 0.008%). It has a density of 8.8, an atomic number 27, and atomic weight 58.94. The name cobalt comes from the German word *Kobold*, meaning goblin, and reflects the difficulty of extracting it from its minerals. Cobalt and nickel minerals often occur together and associated with those of iron and copper. Arsenic is usually present with cobalt in minerals.

Cobalt is found in nature in the minerals cobalt glace ($CoAsS$), linnaeite (Co_3S_4), and smaltite or cobalt speiss ($CoAs_2$). In compounds cobalt exhibits oxidation states of 2+ (cobaltous) and 3+ (cobaltic). The cobaltous ion is quite stable to oxidation, and solutions of cobaltous salts may be exposed indefinitely to the air without change. Most cobaltous solutions are assumed to contain the hydrated ion $Co(H_2O)_6^{2+}$. The cobaltic ion (Co^{3+}) is a powerful oxidizing agent able to oxidize water to form O_2. Only a few cobaltic salts have been made, such as CoF_3 and $Co_2(SO_4)_3 \cdot 18H_2O$.

Cobalt salts commonly employed to correct deficiencies are cobaltous chloride ($CoCl_2 \cdot 6H_2O$), and cobaltous sulfate ($CoSO_4 \cdot 7H_2O$), the former being the chloride preferably used for drenching sheep.

Cobalt Deficiency Disease

In several parts of Australia and New Zealand for about 80 years prior to 1934, raising cattle and sheep had become almost impossible. These areas were usually along the coast. Much of the difficulty was caused by what was described as "coast disease" and "enzootic marasmus" in Australia and "Morton Mains disease" in New Zealand. The usual symptoms of Co deficiency include a gradual loss of appetite, emaciation, rough coat, scaly skin, reproductive failure, and especially anemia. Although sheep could be fattened on the affected coastal lands, they became listless, wasted away, and often died if kept there too long. If, however, the sheep were shifted to the much richer pasture on the adjoining highlands, they invariably recovered their health, but, because of the scanty herbage, failed to gain weight. This indicated that the cause of the sickness was not germs but rather some factor missing in the composition of the herbage. In Scotland a similar disease affecting cattle and sheep called pining, vinquish, or daising was recorded as early as 1831. The disease in Scotland was in time definitely identified with that occurring in Australia and New Zealand.

In 1929 the Division of Animal Nutrition of the Council for Scientific and Industrial Research (Australia) initiated an investigation of the cause and distribution of coast disease in South Australia. Since anemia was one of the symptoms of this disease, large doses of iron salts were tried, but found to be of no value; likewise, treatment with compounds of copper, manganese, and arsenic, alone or with iron salts, had no beneficial effect.

By 1934 the stage was set for conducting a series of critical experiments to find out the cause of coast disease. In Western Australia Filmer and Underwood[4] reported that enzootic marasmus of sheep could be corrected by administering a dose of 0.1–2 mg of $Co(NO_3)_2$ each day, and 2 years later they reported that affected cattle could be benefitted in the same manner[5]. In 1936 Askew and Dixon[6] reported that cobalt cured Morton Mains disease of sheep in New Zealand. That cobalt deficiency might be the cause of the manifestations of the disease received support from spectrographic analyses of soils; two healthy soils contained 7 ppm Co, while soil from a Morton Mains disease area contained no Co at all. Experiments later conducted in Scotland on pining also disclosed the intimate association of cobalt to the disease; it was shown that pining could be cured and prevented for 6 months by giving the sick animal a daily dose of 1 mg Co for 14 days. R. L. Mitchell reported that, in Scotland, most of the soils of which pining was prevalent had a cobalt content of from 1–5 ppm.

In 1958 Beeson and Thacker[7] reported that salt sick, pining, vinquish, coast disease, wasting disease, are only local descriptive names for the one disease affecting sheep and cattle caused by a deficiency of cobalt in the herbage grazed by the ruminants.

Cobalt and Vitamin B_{12}

In 1948 E. Lester Smith in England and E. L. Ricks in the United States discovered that the antipernicious anemia factor of liver extracts

[4] Underwood, E. J., and Filmer, J. F., *Australian Vet. J.*, **11**, 84–91 (1935).
[5] Underwood, E. J., and Filmer, J. F., *Australian Vet. J.*, **13**, 57–64 (1937).
[6] Askew, H. O., and Dixon, J. K., *New Zealand Sci. Technol.*, **18**, 73–92 (1936).
[7] Beeson, K. C., and Thacker, E. J., *Soil Sci.*, **85**, 87–94 (1958).

was an organic compound containing cobalt. It was named vitamin B_{12}.

Vitamin B_{12}, also known as cobalamine, is unique among vitamins in that it contains a metal, cobalt. Its chemical structure is not completely known. The vitamin occurs in all animal tissues, but animals do not synthesize it and must depend on other animals or microorganisms for their supply. Meat is an excellent source of the vitamin. Humans require about one microgram of Co per day for adequate treatment of pernicious anemia. The vitamin accounts partially for the essentiality of sources of animal protein in chick diets, since it is not present in most vegetable products, although the commercial source is fermentation residues from the production of aureomycin. Various microorganisms thus produce it. Manure and sewage are also sources. In cattle and sheep the microorganisms in the rumen synthesize the vitamin, and this makes the cobalt content of the feed eaten by ruminants a very important element in the diet. Cobalt alone will not replace vitamin B_{12}; its principal role to date being a metallic constituent of the vitamin. Specifically what is called cobalt deficiency is vitamin B_{12} deficiency. Hence, it may be observed that all animal life depends upon microbial production.

Function of Cobalt

Although it is now accepted that cobalt deficiency induces pining and that correcting that deficiency by administering the element cures the disease, the precise manner by which it prevents or cures the condition has not as yet been unravelled. What the definite factors are regarding the physiological function of cobalt in the animal body, science has not yet established. As previously indicated, Co has, insofar as is now known, no other function in the physiology of the higher plants than that of supplying the element to the symbiotic *Rhizobium* in the root nodules of legumes for conversion into vitamin B_{12}, and in animal physiology that of furnishing it to the bacteria in the rumen of cattle and sheep to produce vitamin B_{12}. It seems beyond doubt that the value of an adequate supply of cobalt is intimately linked with the presence of this element in vitamin B_{12} and the anemia which afflicts animals suffering from pining is actually caused by an inadequate supply of this vitamin.

Cobalt in Soils

What is meant by soil fertility is a question posed by researchers now that a broader knowledge of trace elements requires a new definition. Is the answer to be based on the end product of the farm, that is, beef, mutton or wool? If so, a cobalt deficient pasture soil, although covered with a good stand of herbage, would be no better than if the soil were completely sterile. If the end product were to be horses or rabbits the deficiency would be of no concern, and the soil would be considered fertile.

It is now considered that herbage containing 0.08–0.1 ppm of cobalt will meet the dietary requirements of cattle and sheep.

McMurtrey and Robinson[8] report that most soils contain from 10–15 ppm Co. R. L. Mitchell[9] found in soils of northeast Scotland a range of from 1–300 ppm Co. He emphasizes that most of the cobalt present in soils is not readily available to plants, being lodged in the crystal lattice of minerals, and that a correlation in comparisons of total soil cobalt with plant uptake is not to be expected.

In Great Britain cobalt deficiency is associated with soils derived from Old Red Sandstone or Devonian rocks, from calcareous shell sands, and from some limestones and granites. Wehrman[10] showed that an inverse relationship exists between the Co content of herbage and the soil pH.

Generally, in deficient soils, the commonly reported low total content is of less than 0.5–3 ppm Co, whereas in soils carrying healthy stock the Co content is usually up to 30 ppm. The availability of the soil Co varies considerably from soil to soil, and meaningful correlations between total content and plant uptake are difficult to obtain.

The availability of Co may be reduced in peat soils. Many forages grown on organic soils have been relatively low in Co. Waterlogged conditions in soil tend to favor availability and plant uptake.

Cobalt in Plants

The essentiality of cobalt as a plant nutrient element has not been established. In the soil-plant-animal relationship, the plant's role is

[8] McMurtrey, J. E., and Robinson, W. O., *U.S. Dept. Agr. Yearbook*, 807–829 (1938).
[9] Mitchell, R. L., *Proc. Nutr. Soc.*, **19**, 148–154 (1960).
[10] Wehrman, J., *Plant and Soil*, **6**, 61–83 (1955).

primarily to accumulate Co for conversion to vitamin B_{12} by microorganisms. Legumes are quite important in this relationship, since they are recognized as very good Co accumulators and in general they contain more than grasses. Andrews gives the following contents as typical of seven pasture species[11] (dry basis):

Plant	ppm
White clover	0.24
Red clover	0.24
Perennial ryegrass	0.16
Short rotation ryegrass	0.13
Meadow fescue	0.12
Cocksfoot	0.11
Timothy	0.09

Price and Hardison found[12] in Virginia grasses that a 3 year average Co content ranged from 0.02–0.24 ppm and of forage legumes, from 0.06–0.43 ppm.

Beeson[13] gives the following data (on dry basis):

	ppm
Field beans (seed)	0.01
Cabbage (edible portion)	0.07
Corn (maize) kernels	0.01
Wheat, grain	0.01
Potatoes	0.06

Corrective Methods

To cure or prevent cobalt deficiency disease, three methods are commonly employed, namely fertilizing the pasture, individual dosing or drenching, and salt licks and mineral supplements. In each of these methods the amount of cobalt required is exceedingly small.

[11] Andrews, E. D., *New Zealand J. Agr.*, **92**, 239–244 (1956).
[12] Price, N. O., and Hardison, W. A., *Va. AES Tech. Bull.*, 165 (1963).
[13] Beeson, K. C., *U.S. Dept. Agr. Misc. Publ.*, No. 369 (1941).

Fertilizing Pastures

Experience in most countries has shown that the most effective method to provide cobalt to grazing animals is through top-dressing the pasture. The length of time during which top-dressings remain effective depends on the frequency of application of fertilizer, the amount of cobalt applied, soil characteristics, and rainfall. Experiments in New Zealand with fairly heavy applications of cobalt have given variable results, but in general, cobaltous sulfate spread from the air to give a rate of 20 oz of the compound/acre will prevent Co deficiency disease on pumice soils for about 3 years. An application of 5 oz/acre of cobaltous sulfate in the autumn of each year is recommended and will increase the Co content of pastures above deficiency levels for a period of 12 months.

Cobaltous sulfate may be mixed with other fertilizers, and it is now general practice to apply it as cobaltized superphosphate, which is produced by dissolving the sulfate in the acid, which is used to treat phosphate rock in the manufacture of normal superphosphate. In the North Island (New Zealand) the processed superphosphate contains 3 lb of cobaltous sulfate/ton, which at annual applications of 2 cwt/acre will furnish the required 5 oz/acre. In the South Island (New Zealand) cobaltized superphosphate contains 6 lb of the sulfate/ton so that annual applications of 1 cwt/acre gives the required Co dosage. If heavier applications of fertilizer are required, such additional quantities do not have to carry cobalt.

Where pregnant ewes are wintered on crops, the preferred method is to apply cobalt to pastures for their lambs during the following spring or early summer, which is just the time when extra cobalt for lambs may be needed. Spraying the pastures with cobalt sulfate solution to give the recommended 5 oz of the crystals/acre is a suitable procedure.

Cobaltous chloride is also used as a source of cobalt. On the average, Australian practice uses 10 lb of this compound (7.9 oz of Co) to treat one acre of pasture.

When considering the role of fertilizers in satisfying the animal requirements for trace elements, agronomists and fertilizer technologists would do well to keep in mind that there are specific differences among them in their functions in the soil-plant-animal chain. Therefore, the fertilizer to be used has to be chosen so that it functions specifically to meet the individual needs of the animal. The system of livestock management involved may be an important factor in deciding whether the animal requirement for micronutrients should be met by fertilizers, feed supplements, or drenching. In the case where animals are fed entirely from pastures, the obvious judgment should be to fertilize the pasture. If the animals are fed grains or other concentrates, adding the micronutrient to the mineral supplement or feed mixtures would be preferable.

Scientists are learning much more about where trace element problems exist both for plants and animals, and also about the place of soil amendments in providing practical means for correcting micronutrient problems. Indiscriminate trace element fertilization will most often fail to meet the needs of either plants or animals. Work by Mitchell and associates[14] indicates the danger that may arise from careless cobalt fertilization; they found by spectrographic analysis that the herbage growing on cobalt treated soil could absorb an abnormally large amount of molybdenum. Where the herbage grew on untreated soil, already supplied with a fairly high Mo content, this amount of Mo might be so increased by the cobalt fertilization as to approach the Mo content of the herbage of teart land, with the consequent scouring among the grazing cattle. This is shown by the data in the following table prepared by Mitchell, et al[14].

Since teartness occurs on soils of pH 7 and above, it would be wise to avoid applying cobalt in the form of a cobalt-rich lime or mixed with fertilizers having an alkaline reaction. Liming causes a decrease in the amount of cobalt plants can take up and may also

TABLE 15.1 Co AND Mo CONTENTS OF HERBAGE FROM COBALT TREATED SOILS*

$CoCl_2 \cdot 6H_2O$ ADDED lb/acre	HERBAGE FROM SOIL A		HERBAGE FROM SOIL B	
	Co ppm	Mo ppm	Co ppm	Mo ppm
0	0.08	1.7	0.07	6.3
2	0.22	2.2	0.20	9.2
10	0.63	2.4	0.89	10.0
80	3.20	7.5	2.75	14.2

* Herbage cut 15 months after various degrees of cobalt treatment.

[14] Mitchell, R. L., et al., *New Zealand J. Soil Tech.*, **36**, 610–613 (1941).

depress the availability of other essential micronutrients, while increasing the availability of molybdenum in mixed herbage 2–3 times.

Drenching

Drenching is a very effective method of getting cobalt into an animal's system by mouth, since an affected animal is thereby assured of receiving the correct dose. The drenching may be done daily or weekly with a stock solution made up by dissolving 1 oz cobaltous chloride crystals in one pint clean, cold water.

For daily drenching the practice is to dilute 1 fluid ounce of the stock solution with 2 pints of clean water; then 1 fluid ounce of this diluted solution is added to $\frac{3}{4}$ pint of water. Each animal is drenched daily with 2 fl oz of the final solution.

For weekly drenching 1 fl oz of stock solution is made up to $4\frac{1}{2}$ pint with clean water, and each animal drenched weekly with 2 fl oz of the diluted solution until completely recovered.

Drenching takes a lot of time and labor and is not practical except in the case of a small flock. Top-dressing the pastures is a means of preventing pining as well as correcting it when present.

Salt Lick

It is best to administer the licks by trough feeding. Licks may be especially suitable in dry climates, but they have the disadvantage that it is impossible to insure individual animals getting the correct dosage; sheep vary a lot in their taste for salt, and it could happen that some would get too much Co and others not enough.

16

IODINE

Iodine (I) is a trace element established scientifically as essential to human and animal life but not to plant life. The public associates iodine with goiter, an affliction characterized by an abnormally enlarged thyroid gland, located in the front of the neck just below the "Adam's apple." It was discovered quite accidentally in 1811 during the British blockade of France, when shipments of Chilean saltpeter, sodium nitrate, used in the manufacture of gunpowder, were cut off. To replace this loss the French set up a nitrification plant, where calcium nitrate was treated with potash-containing ash of kelp to produce potassium nitrate. The nitrate oxidized the potassium iodine in the ash, which volatilized as elemental iodine and crystallized in the vats[1]. This discovery led to investigations of other sources, such as seaweeds, seawater, sponges, marine animals, water cress, and mineral springs.

Iodine is a member of the group of elements collectively known as halogens, which includes bromine, chlorine, and fluorine. The group as a whole has electronegativity, forms negative halide ions as

[1] Gilbert, F. A., "Mineral Nutrition of Plants and Animals," Univ. of Okla. Press, 1948.

they occur in ionic salts, and, except for fluorine, shows positive oxidation states. Each element of the group has seven electrons in the outer ring or energy level. The atomic number of iodine is 53 and its atomic weight, 126.91. At room temperature it is a solid, and it crystallizes in the form of metallic, lustrous, black leaflets. It occurs as ionic I^- in seawater and salt wells and is also found as sodium iodate $(NaIO_3)$. It is only very slightly soluble in water (0.001M). It colors starch blue on contact.

Iodine is not as widely employed as the other halogens. It has antiseptic properties and is commonly used as tincture of iodine (a solution of I_2 in alcohol) and as iodoform (CHI_3). For prophylactic purposes iodine in the form of KI or NaI (0.01%), 0.0076% iodine, is often added in trace amounts to table salt to supply the need for it in the human diet, particularly in those areas where goiter is prevalent.

Iodine in Soils

In "Iodine Facts" the Iodine Educational Bureau (London) published in 1946 a summary of the information on the iodine content of soils available up to that date. It states that, generally, heavy-textured clay soils and loams derived from igneous rocks are richer in iodine than light, sandy soils. The following list summarizes the main characteristics of soils with reference to iodine content[2]:

IODINE RICH	IODINE POOR
Heavy textured.	Light textured, porous, iodine rapidly leached out.
Humus rich.	Low in content of colloidal clay and humus.
Acid in reaction.	Tendency to react alkaline.
Soils having colloidal iron in ferric state.	Contained iron, in ferrous state.
Derived mainly from igneous rocks.	Mainly derived from sedimentary rocks.

In some New Zealand soils iodine content varies widely, and studies show that the incidence of goiter in grazing animals is highest

[2] Corrie, F. E., *Fertilizer J.*, 70 (1948).

on soils with a relatively low iodine content. An example of this is illustrated in the following tabulation of samples from various districts in New Zealand[3]:

DISTRICT	NO. SOIL SAMPLES	MEAN I CONTENT MICROGRAMS PER 100 GRAMS
Auckland	104	530
Wellington	77	480
Canterbury	71	90
Otago	219	130

In Canterbury goiter occurs in many species of animals and to less extent in Otago.

Iodine in Plants

Experimental evidence indicates that the amount of available iodine in a soil has the greater influence in determining the content of iodine in plants than the species to which the plant belongs. Plant species vary in the amount of iodine which they take up, as high in some cases as 5 ppm, but the more common range is 0.2–0.5 ppm, and this difference holds when plants of the same species are grown in different parts of the same district. In culture solutions iodine in the amount of 0.1 ppm was toxic to plants. Brown and Dietz[4] reported that the iodine content of lettuce grown in different parts of Ohio varied widely; leaf lettuce grown in greenhouses in different parts of Ohio showed a range of 1.42–12.5 ppm (dry weight), while samples from the same area had about the same content of iodine.

Orr and Leitch[5], using pasture data from samples in Scotland, estimated that the daily intake of iodine for a 110 lb sheep ranged widely (based on I content of herbage) from 120–2,000 micrograms, and the intake of a 1,100 lb cow ranged from 3,000–30,000 micrograms. They could not record whether the lower levels for each animal species were adequate.

[3] Hercus, C. E., et al., *J. of Hygiene*, **311**, 493 (1931).
[4] Dietz, C., *Ohio Expt. Sta. Bull.*, 592 (1938).
[5] Orr, J. B., and Leitch, I., *Med. Res. Council Spec. Rpt. Ser.*, No. 123 (1929).

The iodine content of plants may be increased by the application to soils of Chilean nitrate of soda, seaweeds, and bird guano, which are materials containing iodine in varying amounts. There is some evidence that organic manures applied to the soil seem to cause the release of iodine already present in the soil in unavailable form, rather than by contributing their own contained iodine[6].

Fraps and Fudge[7] found the iodine content of young Bermuda grass on some soils in Texas lower than that of young Bluestem grass grown on the same soil, and when grown on other soils it was greater.

Calderbank[8] has pointed out that soils and pastures close to the sea are not always richer in iodine than those farther inland. Other factors will influence it, direction of prevailing wind, amount of precipitation, and the nature and pH of the soil.

Seaweeds show a range of iodine content of 40–3,500 ppm. Yields of iodine from European kelp may vary from 12–22 lb/ton of kelp.

Iodine in Animal Nutrition

Beeson[9] reviewed many foreign and domestic reports on the mineral composition of crops and compiled the relevant data. Some of the average figures from his listings of the iodine contents of specified crops (dry matter basis) follow (contents in ppm):

> Barley grain, 0.102; field beans, seed, 0.057; cabbage, edible portion, 0.218; wheat grain, 0.067; sugar beet roots, 0.087; sugar beet tops, 0.425; potato tubers, 0.097; red clover, 0.121.

Iodine is now established as an essential element in nutrition. The Agricultural Research Council (London) in 1966 considered that a level of 0.12 ppm of iodine will satisfy the needs of ruminants except during periods of pregnancy and lactation when the level should be 0.8 ppm. If goitrogens are present (generally present in legumes, especially in white and subterranean clovers), the level should be increased by 1.2 ppm.

[6] McClendon, J. F., "Iodine and Incidence of Goitre," Oxford University Press, 1939.
[7] Fraps, G. S., and Fudge, J. F., *Texas Agr. Expt., Sta. Bull.*, No. 595 (1940).
[8] Calderbank, G., "Iodine in Animal Health, Production and Pasture," London, Longmans, Green and Co., 1963.
[9] Beeson, K. C., *U.S.D.A. Miscel. Pub.*, 369 (1941).

Iodine in Animal Products

Godden[10] gives the following information on the average iodine content of specified animal products:

> Animal flesh, 3–4 micrograms/100 gm fresh material; flesh of fish is generally richer in iodine, the range for saltwater fish being 26–623 micrograms/100 gm fresh material; skimmed milk 7 micrograms/100 gm; refined cod-liver oil, 910–1,650 micrograms/100 gm fresh material; lard, 2–12, butter, 5–14, and vegetable oils 0.4–9.5 micrograms/100 gm.

Iodine and Goiter

Iodine deficiency causes goiter in humans and animals, cretinism in persons, and "hairlessness" in hogs. The development of these conditions has been correlated with certain low levels of iodine in food and water. Iodized salt is used as a preventive. An excess of iodine can produce what is known as "hyperthyroidism," and for this reason physicians recommend that iodine should be taken preferably in the organic form.

George Murray, a British professor of medicine, discovered in 1891 that goiter or thyroid deficiency could be corrected by injecting thyroid extract into the deficient gland. The thyroid gland uses iodine derived from the daily diet to produce a complex chemical compound called thyroxine, which contains about 65% iodine. Thyroxine is the essential, iodine containing secretion or hormone, by whose means the thyroid exerts a profound and regulatory influence on the vital functions. Biologists have discovered that thyroxine is indispensable for physical and mental development, for normal reproductive capacity, for the growth of wool and hair, and particularly for regulating the temperature of the body. It has been said: "The thyroid gland is to life what the draught is to the fire."

Goiter has been reported from almost every country on earth. For a long time the true cause of goiter remained unknown. It is now well established that the cause is iodine deficiency. This is illustrated by the data in the Table 16.1 (as given by Orr and Leitch)[5].

In most cases of goiter in livestock, correction has been achieved

[10] Godden, W., *Agr. Progress*, **8** (1931).

TABLE 16.1 IODINE CONTENT OF RUMINANT THYROIDS IN THE UNITED STATES[5]

| RUMINANT | THYROID | IODINE, gm/100 gm | |
		FRESH WT.	DRY WT.
Cattle	Normal	0.112	0.346
	Enlarged	0.004	0.019
Sheep	Normal	0.069	0.247
	Enlarged	0.0001	0.0006

by feeding iodine to the affected animal and by giving iodine to pregnant animals as a preventive of goiter in the young animal. Hopkirk[11] investigated an outbreak of goiter in lambs in Wanaka, New Zealand and found the pasture low in iodine content (11–17 micrograms I/100 gm, dry weight) compared with 95 μgm for good pasture in the Canterbury District, sampled at the same time. Wanaka pasture was also low in chlorine, sodium, and high in calcium.

Limited data indicate that a low iodine content of pasture and hay may cause goiter, but at times values of iodine content considered adequate in some areas may produce goiter in others; pinpointing the true cause of goiter has been a complicated business.

[11] Hopkirk, C. S. M., et al., *New Zealand J. Agr.*, **40**, 226 (1930).

17

FLUORINE

Fluorine (F), like iodine, is required by animals, but as far as we know now, it is not essential for the normal growth and development of plant life. Fluorine, also like iodine, is always found in minute amounts in soil grown plants, but no one has found that it has any specific function in living matter. Up till now no deficiency symptoms have been recorded in plants and animals subsisting on foods containing the minutest amount of the element. It becomes important from the viewpoint that, in animal nutrition, it may be harmful even when ingested in minor doses.

Fluorine Sources

It is widely distributed in rocks, soils, and waters particularly associated with phosphorus in phosphate rock as fluorapatite $(Ca_5(PO_4)_3F)$, and with calcium as calcium fluoride (fluorspar), and in nature as cryolite, an impure double fluoride of sodium and aluminum (Na_3AlF_6). It also occurs in combination with silicates as topaz, tourmaline, and the micas. Goldschmidt[1] estimated that the earth's crust contains per ton,

[1] Goldschmidt, V. M., *J. Chem. Soc.*, 655–673 (1937).

270 gm of fluorine, and 786 gm of phosphorus, with a ratio of F : P of about 0.34 to 1, about the same ratio as in fluorapatites. Fluorine is present in all United States phosphate rock deposits, the amount ranging from 2.65–3.90% F.

Fluorine is widely distributed in drinking water in the United States and throughout the world. In many American communities fluorine is added in controlled amounts (1 ppm) to city water supplies as a means of strengthening the structure of the teeth, especially children's, to reduce caries. Fluorine is beneficial to some animal life, at least those that have bone skeletons, for fluorine is an essential constituent of apatite, the calcium phosphate mineral of which bones and teeth are composed. Skeletons cannot develop normally without minute amounts of fluorine. Amounts of fluorine above 2 ppm in drinking waters are considered excessive and can weaken the crystalline structure of the teeth's enamel. Experience with animals' drinking water containing 10–20 ppm fluorine indicates that the tooth enamel was softened to a degree where rapid erosion resulted and the animals could no longer chew their feed.

Fluorine in Plants

Fluorine is usually present in plants in the range of 2–20 ppm F dry matter. Tea is one of the best known accumulators of the element; its dried leaves may contain as much as 400 ppm[2]. No evidence, however, exists to indicate that fluorine performs any essential function in plant nutrition. Fertilization of the soil with phosphates and slags carrying fluorine in their composition does not apparently increase the fluorine content of plants benefitting from such fertilization. The incorporation of soluble fluorides into an acid soil may be toxic to seedlings or may prevent seed germination, but only when the concentration reaches a high level[3]. Such accumulation of soluble fluorides may be neutralized by adding liming materials or soluble phosphates to the soil; the phosphate tends to produce the insoluble fluorapatite. Fluorine containing insecticides have been identified as a cause of toxicity to plants. Most of the fluorine toxicity to plants, however, has been caused by the atmosphere, particularly near factories which process phosphates

[2] Bear, F. E., *USDA Yearbook*, 167 (1957).
[3] Morse, H. H., *Soil Sci.*, **39**, 177–195 (1935).

and aluminum. Thermal processing of phosphate rock, which averages 3–4% fluorine, causes about 90% of this fluorine to be volatilized in the form of silicon fluoride. When this fluoride contacts moist air, it yields hydrofluoric acid (HF) gas, which is quite toxic to plant foliage.

Fluorine loses its toxicity when "fixed" by vegetation, such as by alfalfa.

Fluorine in Animal Nutrition

Animals have a limited requirement for fluorine in the diet, at least the vertebrates do, because it is a component of teeth and bones. In proper amounts it helps to reduce the tendency to tooth decay, but an excessive amount can lead to a rapid breakdown of the enamel surface of the teeth[4].

It is estimated that a person ingests daily about 0.75 mg F without any suspicion of harming health or nutrition. If more than this amount is taken in, the excess is excreted mainly in the urine. The only injury possible from a moderate excess is, as previously noted, to the teeth and bones.

Fluorosis

Fluorosis or fluorine poisoning in humans is generally caused by eating or inhaling fluorine containing gases and dust; in animals by the ingestion of apatitic supplements to low phosphorus rations. The amounts of fluorine normally present in animal tissues vary from less than 1 ppm for blood to about 250 ppm for bone (dry basis). The symptoms of chronic and acute fluorosis in man and animals have been reviewed by Roholm[5], Greenwood[6], and by Pierce[7].

Experience gained with farm and laboratory animals indicates that a safe concentration of fluorine in the total ration of animals is 100 ppm, dry basis, or a little less. For chickens, the level may be

[4] Deatheradge, C. F., *J. Dental Res.*, **22**, 173–180 (1943).
[5] Roholm, K., Fluorine intoxication, Copenhagen and London (1937).
[6] Greenwood, D. A., Fluoride intoxication, *Physiol. Rev.*, **20**, 582–615 (1940).
[7] Pierce, A. W., *Nutr. Abstr. Rev.*, **9**, 253 (1939).

somewhat higher. For man the amount of about 50–100 ppm F, dry food basis, is safe[8].

Defluorinated phosphate rock is prepared in the United States to compete with or replace bone meal or bone flour. It apparently is a good source of available phosphorus used as a supplement to the ration of farm animals and poultry.

[8] Mitchell, H. H., National Res. Council, Reprint and Cir. Ser., Rpt. 10 (1942).

18

NICKEL, LITHIUM, VANADIUM, SILICON, ALUMINUM

Nickel

Nickel (Ni) is an element poisonous to plants even at relatively low concentrations. Analysis shows that soils derived from serpentine rocks are unusually high in nickel content. Total nickel in soils generally ranges between 10–40 ppm. The nickel content of plants averages between 0.5–2 ppm. Experiments at Rutgers University showed that a level of 40 ppm Ni in tomato plants was toxic, and a level of 150 ppm stopped their growth[1]. Mitchell[2] states that the normal nickel content of plant material, dry basis, ranges from 0.10–5 ppm, depending on the species, the part of the plant, maturity, time of sampling, and other factors. He recorded a level of 0.5–4 ppm Ni in pasture herbage sampled in northeast Scotland.

[1] Bear, F. E., *U.S. Dept. Agr. Yearbook* (1957).
[2] Mitchell, R. L., *Soil Sci.*, **60**, 63–70 (1945).

Nickel is not considered essential to either plant or animal life. Roach and Barclay[3], however, found statistically significant increases in crop yields in their Romney March experiments following the application of nickel sprays. Apparently an element that by the pure solution culture technique is found to be nonessential to the life and growth of a plant may still be beneficial.

As for its use in animals, some work in Australia and New Zealand indicates that nickel could replace cobalt in the treatment of "enzootic marasmus" and Morton Mains diseases.

Lithium

Lithium (Li), often associated in small amounts with sodium and potassium in the mica-type minerals, produces a characteristic mottling

TABLE 18.1   LITHIUM CONTENT OF CITRUS LEAF SAMPLES FROM SOUTHERN CALIFORNIA[4]

CITRUS GROWING AREA	NO. OF SAMPLES	LITHIUM CONTENT ppm	CITRUS VARIETIES	DEGREE OF VISUAL TOXICITY SYMPTOMS
Coachella Valley	25	5–40	Grapefruit, lemon, tangerine, Valencia, and navel oranges	Moderate–severe
Santa Barbara Co.	10	2–15	Grapefruit, lemon, tangerine, Valencia, and navel oranges	Moderate–severe
Ventura Co.	5	8–15	Valencia oranges, lemons	Moderate–severe
Riverside Co. (Hemet)	4	5–20	Grapefruit, navel oranges	Slight–moderate
San Fernando Valley	3	5–25	Valencia oranges, lemons	Slight–moderate
Tulare County	3	5–15	Navel oranges	Slight
Imperial Valley (Holtville)	4	25–35	Valencia oranges	Moderate–severe
Desert areas (Desert Center and Vidal)	8	30–40	Grapefruit, lemon, orange, mandarin	Moderate–severe

Source: *California Agr.* (December, 1961).

[3] Roach, W. A., and Barclay, C., *Nature*, **157**, 696 (1946).

on citrus leaves when present in small amounts in the soil or irrigation water. Such mottling was detected in 62 leaf samples representing eight citrus growing areas of southern California[4].

The results of a survey for trace elements in 43 high producing citrus groves in southern California showed that lithium was present in amounts ranging from 5–22 ppm in about 15% of the samples; the amount in the remaining 85% is not given. The following table summarizes the results of the survey.

Vanadium

In a brief notice in *Nature*[5], Arnon and Wessel published a report which seemed to show quite definitely that the element vanadium (V) is a beneficial, if not essential, nutrient to at least one species of green plant, the alga *Scenedesmus obliguus*. Using an iron salt in their study of iron nutrition effects on this alga, they obtained such exceptional increases of growth with increases in iron concentration that they were inclined to suspect an impurity in the iron salt they were using. This salt was in fact not 100% pure. The most likely impurities would be cobalt, vanadium, or nickel. They then used a very pure ferric chloride and added trace amounts of salts of these three suspects, and by a process of trial and exclusion they found that vanadium was the factor responsible for the increase in growth.

Vanadium (V) is a fairly widely distributed chemical element, forming 0.032% of the 10 mile depth of the lithosphere. Vanadium has an atomic number of 23 and atomic weight of 50.95. It is more abundant than copper or zinc. Reports of plant analyses have indicated that vanadium was present in 62 plant species out of 62 analyzed, but physiologists have not been willing to accept presence as proof of essentiality. Bortels[6] reported that vanadium was one of the elements that promoted growth and nitrogen fixation in the azobacter organism, and he assumed it acted as a catalyzer.

Silicon

Silicon (Si) is considered an element beneficial to plant growth but not indispensable to it, nor has it been successfully demonstrated to be

[4] Bradford, G. R., *California Agr.* (December, 1961).
[5] Arnon, D. I., and Wessel, G., *Nature*, **173**, 1039–1040 (1953).
[6] Bortels, H., *Arch. Mikrobiol.*, **1**, 333–342 (1930).

essential to animal life. Silicon is the second most abundant element, 26%, in the earth's crust, where it is present in the form of quartz and silicates. Together with aluminum oxide, ferric oxide, and water, it makes up the main bulk of the soil colloids or clay. It is about as important in the mineral world as carbon is in the organic. As silica (SiO_2), it accounts for sand, flint, quartz, and opal. It has an atomic number of 14 and atomic weight of 28.09.

Early investigators considered silicon essential for plants, since it constitutes a large part of their makeup. Later research showed that it may be essential only to barley, sunflower, and beets[7]. Some workers have shown that a possible relationship exists between silicon and phosphorus metabolism[8]. Since 1934 Japanese scientists have considered silicon essential for the normal growth of rice[9]. Other Japanese research shows that deficiency of Si in rice plants lowers resistance to diseases and insects.

Silica is present in appreciable amounts in many plants. In the case of beet it is said to be indispensable; in rice and barley it is important as a nutrient, and a host of other crop plants seem to benefit from it. Its precise function in the plant remains a question for future investigators to determine. So far, considerable support favors the view that silica serves not only to nourish the plant but also to provide resistance to fungal disease and to insects. Lanning[10], however, studied the relation of silicon in barley to resistance to disease, pests, and to cold. He found no direct relationship between total silica content and resistance to greenbugs, cold, or diseases. The matter calls for more study that may clarify the contradictory results so far reported.

Taylor[11] published a review of the literature about silicon and silicate effects on making soil phosphates more available to plants. The evidence from agronomic research indicates that silicates as fertilizers do not have economic value. Taylor is quite doubtful about the suggestion that silica is valuable because it can displace phosphorus from locked up combinations with iron and aluminum, because the silicate benefits are obtained at the expense of the soil's phosphorus reserves. Ordinary crystalline silica, finely ground, has failed, when tested agronomically as a phosphate releaser, although at an earlier

[7] Raleigh, G. J., *Plant Physiol.*, **14**, 823–826 (1939).
[8] Nemec, A., *Biochem. Z.*, **190**, 42–56 (1927).
[9] Oota, M., *Bull. Fac. Lib. Arts and Educ.*, Yamanashi Univ., **5**, 183 (1954).
[10] Lanning, F. C., *J. Agr. Food Chem.*, **6**, 636–638 (1966).
[11] Taylor, A. W., *J. Agr. Food Chem.*, **9**, 163–165 (1961).

date German field tests had shown that cereal crops responded favorably when soluble silicates had been applied as fertilizer. Another question has been raised regarding silica as a fertilizer; if silica is needed as a beneficial plant nutrient, is it ever necessary to apply it, if soils contain so much of it in an available form.

Silicon in Animals

Silicon is present in all animal tissues, especially connective tissue. Maynard[12] reported blood serum of farm animals contains 1–2 mg Si/100 ml. Fearon[13] states that higher animals average from 0.047–0.164 mg Si/gm of fresh material. Silicon makes up a large part of the ash of feathers, and it is believed it greatly helps to maintain their rigidity.

Although silicon is not known to be toxic, cases are on record where cattle have died as a result of lacerations of the walls of the digestive organ caused by the sharp siliceous spikes of rice hulls[14].

Aluminum

Sommer[15] carried out critical water-culture experiments to test whether aluminum (Al) was essential to plants. Peas and millet were the test plants. Special highly purified salts were employed for the culture solutions, and in those containing aluminum the element was present in the amount of 1 ppm of the solution. The aluminum slightly increased the dry weight of the peas but caused a marked increase in the weight of the millet seed, namely 4.98 gm with and 0.23 gm without the aluminum.

Lipman[16] reported aluminum gave similar increases in sunflowers and corn (maize), particularly in corn yields; the addition of aluminum in the sulfate form at the rate of 1 ppm Al, renewed at certain intervals, increased the dry weight of the vegetative parts by about 20% and of the ears by about 155%. Lipman regarded the results of his tests as

[12] Maynard, L. A., "Animal Nutrition," New York, McGraw-Hill Book Co., 1937.
[13] Fearon, W. R., "Introduction to Biochemistry," London, W. Heinemann Ltd., 1946.
[14] McMurtrey, J. E., and Robinson, W. O., USDA Yearbook, 807–829 (1938).
[15] Sommer, A. L., *Univ. of Calif. Publ. Agr. Sci.*, **5**, 57–81 (1926).
[16] Lipman, C. B., *Soil Sci.*, **45**, 189–198 (1938).

proof that aluminum was essential but was not too certain that his control cultures without the element were in fact absolutely free of it. It is a known fact that silicon and aluminum are always found in soil-grown plants, in some instances to the extent of half their total ash constituents. Most of the bulk of mineral soils consists of aluminum silicate minerals. Soil grown plants always take up these two elements. The presence of silicon and aluminum in the ash of plants illustrates the general proposition that plants absorb soil materials in solution, whether they need them in their physiological processes or not. Evidence exists in support of the claim that aluminum in trace amounts may be essential to some plants, but since most soils contain substantially large amounts of total aluminum, it is most unlikely that a deficiency will ever develop.

Soluble aluminum in concentrations of at least 10–20 ppm may be toxic to crop plants. This effect may be associated with soil acidity. Magistad[17] showed that aluminum solubility in the soil solution at various reactions parallels the curve for the solubility of aluminum in water at the same reactions. As the acidity of soil solutions decreases to the pH 7 level, the solubility of aluminum similarly decreases to almost zero; and when the acidity drops below pH 5, the solubility of aluminum increases fairly rapidly up to pH 4.5, when it increases very quickly with increases in degree of acidity.

Aluminum is often used as a soil acidulant to create an acid soil environment for acid loving plants such as azaleas, rhododendrons, and mountain laurel. The resulting acidity is explained on the basis that the aluminum ion replaces calcium and thereby increases the soil pH. Florists use aluminum to produce blue flowers in *Hydrangea*; the color is produced by the formation of a complex involving aluminum and delphinidine diglycoside[18].

Aluminum is not a constituent of either plant or animal matter, declared McCollum and associates[19]. Other investigators have shown that some mammalian tissues may contain, on average, about 0.5 mg Al/kg and, furthermore, present evidence that involves aluminum in the functions of certain catalysts in the mammalian body[20].

[17] Magistad, O. C., *Soil Sci.*, **20**, 181–225 (1925).
[18] Chenery, E. M., *J. Roy. Hort. Soc.*, **62**, 304–320 (1937).
[19] McCollum, E. V., et al., *J. Biol. Chem.*, **77**, 753–768 (1938).
[20] Horecker, B. L., et al., *J. Biol. Chem.*, **128**, 251–256 (1939).

FERTILIZER
AND MARKETING
PROBLEMS

19

TRACE ELEMENTS
IN FERTILIZERS

Although the problem of trace element deficiencies promises new markets for the fertilizer manufacturer, there are difficulties. The production of mixed fertilizers fortified with trace elements is officially recommended, and, on the basis of individual farm, soil, and plant tests, particularly poses many problems in manufacture, marketing, and application. Some trace elements are reported deficient throughout a wide area, others in restricted areas. To learn how to alleviate these situations effectively continues to engage the attention and research efforts of federal, state, and private industry agencies.

The use of mixtures containing several trace elements is a comparatively new practice in this country, and, consequently, much has to be learned on proper methods of mixing, storing, and handling them. Tailor mixing a fertilizer containing one or more trace elements for a single farm is frequently necessary, because such a fertilizer cannot be used indiscriminately but only for the prescribed farm; applying an excess amount of a trace element on a crop with a narrow tolerance limit or in a soil in which the deficiency is minor can damage other crops grown in rotation on the same field. Hence, the manufacturer must use separate bins and equipment for the production of fertilizer-trace element formulations and to carefully avoid contamination of regular grade fertilizers with trace element carriers. In some parts of the country it is possible to prepare and store a relatively large tonnage of a fortified grade fertilizer in separate bins or in bags in anticipation of sales. For example, Wisconsin recommends a borated, top-dressing grade such as 0-13-20 for alfalfa. In that state it is reported that about 60% of its 2.5 million acres of alfalfa is deficient in boron. The official recommendation is to top-dress with a fertilizer containing about 10% borax or borate. Since alfalfa has a high tolerance for boron and, most other crop plants do not, it is possible that a high residuum of boron in the soil might be lethal to a sensitive crop that may follow it in rotation. But such large areas of a known deficiency as in Wisconsin are not common, and elsewhere the manufacturer faces production, sales, and distribution problems.

Another major consideration facing the manufacturer is that the addition of the trace element carrier to a standard grade dilutes it, in some cases significantly, and this means the fortified fertilizer has to be registered as a new grade. The alternative is to formulate a new grade to accommodate the added micronutrient, in which the N, P_2O_5, and K_2O contents are upgraded. In either case it means a separate registration and fee.

The experience of manufacturers in all sections of the country confirms the general difficulty of attempting to adopt a specific quantity of a micronutrient for addition to a standard grade. The reason is obvious; the recommended rates of application of the mixed fertilizer vary widely, from farm to farm, and, frequently, for the same crops. Today, this situation is aggravated by the necessity to comply with the indications of a soil and possibly a plant test. Furthermore, the need for one or more trace elements and their response when applied will be greatly influenced by soil type, its pH, and prevailing climatic conditions. For these reasons it becomes incumbent on the manufacturer that he prepare a "prescription type" of fertilizer mixture, which conforms with the specific results of the soil test his farmer customer has obtained, rather than offer a standardized general type of NPK trace element fertilizer.

The extent and intensity of the trace element problem in any one area or region will influence the fertilizer manufacturer's policy of whether to incorporate trace element compounds into his regular grades by prescription or on an average one-rate basis. In those parts of the country where the soil has been cropped for many generations, as in the Atlantic seaboard region, the extent of trace element deficiencies is much more serious than in the Southwest or Pacific Regions, where much of the cultivated soil was handled at least for the first 30 years or so as a dry farming operation.

Iron is the element most commonly deficient in the southern states, but the local fertilizer industry has not tried to incorporate inorganic sources into the regular grades. Recently, however, chelated iron products have been used with much success in correcting the deficiency. Zinc deficiency is also reported from the South, particularly from the Gulf area east of the Mississippi. Many soils in Florida and parts of the Midwest report deficiencies involving all the essential micronutrients, but, of course, not always on the same farm or in the same community.

The early research on trace element deficiencies in Florida, originating in the late 1920's, greatly stimulated interest in these elements in scientific circles here and overseas. Once established, the fertilizer industry had to learn the best way to market the trace elements. With time and patient trial and error experience, the industry gained confidence in their sales programs. One of the pioneer companies promoted a combination of trace elements, a sort of shotgun mixture of the elements, formulated according to the formula submitted by the customer. This practice is not recommended, since it is not specific enough to meet present day demands. The "shotgun mixture" application involves the risk of upsetting the balance of plant nutrients in the soil. If no problem actually exists and the application is made as a form of insurance against a possible development of deficiencies, the action is useless and could be harmful where the problem is imbalance or toxicity. What really needs to be done is, first, to establish that a deficiency does in fact exist, and then to determine the effective rate to apply on the basis of soil and plant tests. This information should then govern the formulation of the complete fertilizer-macro- and micronutrients.

20

PRIME PRODUCERS
AND MARKETING

Trace Element Carriers

During the past decade numerous purveyors of trace element carriers
have appeared in the market place. Each offers his product in a specialized form with various claims to attract attention and earn a share of
the market. Many compounds are classified as completely water-
soluble, some are moderately to slightly water-soluble, others are
sequestered or chelated. The inorganic sulfate forms are generally
completely water-soluble, most economical, and found in most
markets as solids (powder or granule forms) for use in solid or liquid
fertilizers. They have the disadvantage of being likely to leach in
open, sandy soils. The chelated materials seem to be the most efficient
type of material. They cost more than the inorganic type of soluble
materials. When, however, they are compared on the basis of effectiveness, they appear quite competitive on a per acre, net cost basis. The
chelates are rated at a 5 to 1 effectiveness, that is, only about one fifth
as much is needed as a water-soluble, inorganic carrier to do the job.
Control of iron deficiency in Florida citrus was the first outstanding

success achieved by means of chelated iron. Its superiority lay in its ability to facilitate the translocation of iron throughout the plant, in contrast to the effect produced by the inorganic salts. Chelated trace elements are frequently employed to furnish higher concentrations of the elements in liquid mixed fertilizers, because they are not so prone to engage in chemical reaction which the liquid conditions favor. Another advantage to their credit is their resistance to reactions in highly alkaline soils.

Fritted trace elements and the polyflavonoid series are classed as the slowly available type of carriers.

Boron is generally used as the borate 46% B_2O_3 grade, containing 14.3% B. It is also available in the 65–66% grade containing 20.2% B.

Copper is used in the sulfate form, commonly as $CuSO_4 \cdot 5H_2O$, containing 25.2% Cu, or as the oxide (CuO), containing 70% Cu. Basic copper sulfate, 53% Cu, is used principally for spray applications.

Iron is obtainable as the sulfate and in chelated, fritted, and polyflavonated forms.

Manganese is marketed as sulfate and oxide, with the oxide (MnO) being generally recommended. The manganic form (MnO_2) is also available, and although it costs less it also is less available nutritionally. Manganous MnO is sold in two grades, 48% Mn and 60–65% Mn, respectively, and is marketed both in powder and granular forms.

Molybdenum is available as sodium or ammonium molybdate, as molybdic acid or molybdic trioxide.

Zinc is generally preferred in the sulfate form ($ZnSO_4 \cdot H_2O$), containing 36% Zn, and is either powdered or granulated. The oxide (ZnO), 78–80% Zn, is less soluble than the sulfate form.

For agricultural purposes it is entirely satisfactory to use these trace element carriers in the technical or fertilizer grades. They are also prepared as pharmaceutical or C. P. (chemically pure) grades, but such higher purity is not required and would cost much more than the grade for fertilizer or spray purposes.

The number of suppliers of these micronutrient materials is large and growing. The number of primary producers is comparatively small. The following list, prepared by H. Gordon Cunningham[1], is limited to those American basic producers marketing the technical or fertilizer grade materials.

[1] *Farm Chem.* (March, 1966).

PRIMARY PRODUCERS OF MICRONUTRIENTS

Borate
American Potash and Chemical
Corp.
Stauffer Chemical Co.
U.S. Borax and Chemical Corp.

Copper oxide
Calumet Division, Calumet
and Hecla, Inc.

Copper sulfate
Mountain Copper Co.
Phelps Dodge Refining Corp.
Tennessee Corp.

Ferric sulfate
Stauffer Chemical Co.
Tennessee Corp.

Ferrous sulfate
E. I. duPont de Nemours and
Co., Inc.
U.S. Steel Corp.

Ferrous ammonium sulfate
Kerley Chemical Co.

Iron chelates
Dow Chemical Co.
Geigy Agricultural Chemicals,
Division of Geigy Chemical
Corp.
Hampshire Chemical Corp.
Tennessee Corp.

Iron oxide
Prince Manufacturing Co.
Tamms Industries Co.
C. K. Williams and Co.

Manganese sulfate
Carus Chemical Co., Inc.
Eastman Chemical Products, Inc.
Manganese Chemicals Corp.
Tennessee Corp.

Manganous oxide
E. J. Lavino and Co.
Manganese Chemicals Corp.
Prince Manufacturing Co.
Tennessee Corp.

Molybdenum
Climax Molybdenum Co.
North Metal and Chemical Co.
S. W. Shattuck Chemical Co.

Zinc oxide
American Zinc, Lead and
Smelting Co.
New Jersey Zinc Co.
Prince Manufacturing Co.
Sherwin-Williams Co.
Tennessee Corp.

Zinc sulfate
Chemical and Pigment Co.
Eagle Picher Co.
Polymet Co.
Sherwin-Williams Co.
Tennessee Corp.
Virginia Chemicals and
Smelting Co.

Chelates other than iron
Davies Nitrate Co.
Dow Chemical Co.
Geigy Agricultural Chemicals,
Division of Geigy Chemical
Corp.
Hampshire Chemical Corp.
Rayonier, Inc.

MICRONUTRIENT COMBINATIONS

COMPANY	TRADE NAME	INGREDIENTS
Bennett Chemical Co.	Zinc-El-Izer	Zn, Fe
Consolidated Mining and Smelting Co., Inc.	ZMNS	Zn, Mn, S
Davies Nitrate Co.	Nutramin	Mn, Fe, Cu, B, Mo, Zn
Ferro Corp.	F-T-E	B, Fe, Mn, Cu, Zn, Mo
Kerley Chemical Co.	13-13-13	Zn, Fe, S
Tennessee Corp.	Es-Min-El	B, Cu, Fe, Mn, Zn, Mg
	1000 Mix	Same
	Tenam	Fe, Cu, Mn, Zn, B.
	USP No. 1	Mn, Cu, Zn, B, Mg
	USP No. 2	Zn, Mn, Cu, B, Fe, Mg
	USP No. 3	Fe, Mn, Zn

Size of Market

At present official statistics are lacking regarding the annual United States consumption of trace elements for agricultural purposes. Any estimate of such annual consumption has to be more or less guesswork. The U.S. Department of Agriculture does, however, release data on the magnitude of annual direct sale to farmers of some of the elements.

For the fertilizer year 1966–1967, sales were as follows (in short tons):

Aluminum compounds	104
Boron compounds	3,431
Copper compounds	722
Iron compounds	16,740
Manganese compounds	5,728
Zinc compounds	10,277
Mineral mixtures	10,277

The above figure on boron compounds does not reflect the total amount used. Most of the application is made on the alfalfa crop, which by rough estimation involves about 26 million acres. If, to be conservative, it is estimated that only half the acreage received 2 lb/acre/year of borate equivalent, the total amount would be some 13,000 tons. Some other crops also received borate applications which amounted to about 2,000 tons/year. This means consumption of boron compounds would total some 15,000 short tons, and this figure is on the conservative side.

Serviss[2] estimates that the total amount of zinc compounds, used mainly on the corn crop, would be close to about 36,000 tons (9,000 tons Zn).

Compared with the U.S. total tonnage of fertilizers consumed per year, the tonnage of trace elements is quite small. Many companies, however, both large and small, are beginning to realize they will have to supply the growing demand for micronutrient compounds that research and farm experience are generating in all sections of the vast U.S. agricultural areas. The market for trace elements is bound to grow in volume from year to year. It will remain a specialty market, profitable if well established by knowledge, patience, and honest services. By acknowledging that trace element deficiencies are specific and not general, the manufacturers and distributors will attain their sales and profit goals; such a policy is a must.

[2] Serviss, G. H., *Agr. Chem.*, **7**, 63 (1967).

21

ANALYSIS BY ATOMIC ABSORPTION

Atomic absorption is being rapidly adopted by private and governmental laboratories as a quick and accurate analytical technique to determine the metallic elements in plants and soils. The earliest workers who applied the atomic absorption method to chemical analysis were agricultural chemists working in New Zealand[1]. The method provides a means of analyzing many metals quantitatively at the 1 ppm level: Zn, Cu, Mn, Fe, Co, Ni, Ca, Mg, K, Na. The technique is free of radioactive and other interferences that result from temperature changes.

Determination of the four metallic trace elements, copper, manganese, zinc, and molybdenum, is performed at many state regulatory laboratories by a Model 330 Perkin-Elmer Atomic Absorption Spectrophotometer. Boron is analyzed by standard hot water extraction and subsequent colorimetric determination. The Model 303 instrument detects metallic elements in all soluble materials down to 1 ppm level. In some respects the technique resembles flame photometry in that the

[1] Allan, J. E., *Analyst*, **83**, 466 (1958).

sample material is flame-vaporized for analysis. In atomic absorption, however, the flame is irradiated by light from a hollow cathode lamp with a cathode made of a metal identical to the metal being determined. The amount of light absorbed gives a direct indication of the amount of the metal that is present.

Colorado Soil Test for Zn and Fe

Agronomists at Colorado State University developed a new soil test which diagnoses both zinc and iron deficiencies. The method consists of shaking a few grams of soil with a buffered solution containing DTPA (diethylenetriamine pentaacetic acid). This chemical acts as a mild chelating agent, which extracts the easily soluble zinc and iron. The dissolved elements are then measured by the atomic absorption spectrophotometer.

The extracting solution is buffered at pH 7.3 by triethanolamine and in addition includes calcium chloride to prevent the dissolution of calcium carbonate. These conditions permit the right amount of zinc and iron to be dissolved to indicate their availability in the soil. The test is suitable for neutral and calcareous soils.

Besides the atomic absorption spectrophotometer, some special soil preparation equipment is also required in order to prevent contamination of the soil sample with zinc and iron.

Analysis of Fritted Nutrients

Clemson University, Department of Agricultural Chemical Services, released the results of a study involving the analysis of fritted micronutrients by the atomic absorption technique; Mn, Cu, Zn, and Fe were determined on 17 random samples of commercial fertilizers from five suppliers[3]. The procedure used is as follows:

> One gram of fertilizer sample, placed in a 250 ml Florence flask, was digested in 10 ml of concentrated HCl on a hot plate at a slow boil, until nearly dry. Then, 20 ml of 0.5 N HCl were added. The mixture, boiled for a few minutes, was filtered into a 100 ml volumetric flask. The residue was thoroughly washed with hot distilled water and

[2] David, D. J., *Analyst*, **83**, 655 (1958).
[3] Hammar, H. E., and Page, N. R., *Atomic Absorption Newsletter*, **6**, No. 2 (1967).

washings were added to the flask. The mixture was diluted to volume with distilled water. Analysis was performed on a Perkin-Elmer 303 atomic absorption spectrophotometer. Boron was determined by electrometric titration[4].

Three digestion procedures for dissolving the fertilizer samples were compared: $KHSO_4$ fusion, solution in a mixture of HF-HNO_3, and solution in concentrated HCl alone. The $KHSO_4$ fusion gave lower values for copper and iron than either HF-HNO_3 or HCl digestion. Table 21.1 shows average percentages of manganese, copper, zinc, and iron found in duplicate determinations on six fertilizer samples containing fritted micronutrients for each dissolution procedure. All samples reported in the final results were treated by the HCl method.

Values are mean of duplicate determinations for six fertilizers containing fritted micronutrients.

Means for each element followed by different letters are different at 5% level (Duncan's multiple range)[5].

The conventional, official wet chemical methods, and petrographic and spectrographic emission techniques can all be used for the qualitative and quantitative analysis of soil but are not as sensitive as the atomic absorption technique, especially for zinc. The atomic absorption technique permits a rapid, simple analysis of a single sample for many metallic elements; all that is required is merely the change of lamps on the Perkin-Elmer instrument, and a single sample may be analyzed for many different metallic elements. The savings in time and cost are obvious. The usual methods are based on single element

TABLE 21.1   INFLUENCE OF THREE DISSOLUTION METHODS ON PERCENTAGE OF MANGANESE, COPPER, ZINC, AND IRON FOUND IN FERTILIZERS

DISSOLUTION METHOD	% MANGANESE	% COPPER	% ZINC	% IRON
$KHSO_4$ fusion	0.0608 (a)	0.0500 (a)	0.0722 (a)	0.6107 (a)
HF-HNO_3 digestion	0.0622 (a)	0.0610 (b)	0.0729 (a)	0.6597 (b)
HCl digestion	0.0625 (a)	0.0602 (b)	0.0721 (a)	0.6572 (b)

[4] "A.O.A.C. Official Method of Analysis," 10th Edition, 25, 1955.
[5] Duncan, D. B., *Biometrics*, **11**, 1 (1955).

analysis. It is important to know the status of all the important metallic trace elements in a soil sample, since plants depend on their proper balance. As previously noted, the excess or deficiency of a single micronutrient can frequently be the result of an imbalance of other elements.

The development of the atomic absorption spectrophotometric technique must be regarded as a very important contribution to the analytical tools of the agricultural chemist. It is now possible to make precise recommendations for adding trace elements in soils of all farming areas. It is an indispensable modern tool in scientific agriculture.

References for Further Reading

(1) Allan, J. E., "The Determination of Zinc in Agricultural Materials by Atomic Absorption Spectroscopy," *Analyst*, **86**, 530 (1961).

(2) Allan, J. E., "The Determination of Copper by Atomic Absorption Spectrophotometry," *Spectrochim. Acta*, **17**, 459 (1961).

(3) Gatehouse, B. M., and Walsh, A., "Analysis of Metallic Samples by Atomic Absorption Spectroscopy," *Spectrochim. Acta*, **16**, 602 (1960).

Appendix

Micronutrient Deficiencies

According to a survey reported by Page[1], a deficiency of one or more micronutrients has been found in every one of the states of our nation. This is summarized in the following table, which also gives an estimate of the deficiency acreages and the range in lb/acre of the element recommended to correct the deficiency.

TABLE A.1   MICRONUTRIENT DEFICIENCY IN THE U.S. AND
            RATES OF ELEMENT TO CORRECT IT

ELEMENT	NO. OF STATES REPORTING	DEFICIENCY AREAS MILLION ACRES*	RATES TO APPLY lb/acre†
Molybdenum	27	1.5	0.02–0.4
Boron	44	12.0	0.25–10.0
Zinc	43	6.5	0.35–20.0
Iron	25	3.8	0.5 –10.0
Copper	14	0.6	1.2 –25.0
Manganese	30	13.0	0.3 –20.0

* Date by W. D. Burgess, Jr., Allied Chemical Co.
† Element basis.

[1] Page, N. R., *Croplife*, **2** (1967).

237

Conversion Factors

Different ways to express mineral composition

1 milligram = 1,000 micrograms
10,00 milligrams = 1 gram (1 gm)
1,000 grams = 1 kilogram (1 kg)

1 microgram per 100 grams
= 10 micrograms per kilogram
= 0.01 part per million (ppm)
= 0.000001 per cent

To convert micrograms per 100 grams to parts per million, multiply by 0.01 or divide by 100. Conversely, multiply by 100.

To convert micrograms per kilogram to parts per million, multiply by 0.001 or divide by 1,000. Conversely, multiply by 1,000.

To convert parts per million to percentage, multiply by 0.0001 or divide by 10,000. Conversely, multiply by 10,000.

COMPARISON OF PERCENT AND PARTS PER MILLION

PERCENT (PARTS PER HUNDRED)	PARTS PER MILLION
0.000001	0.01
0.00001	0.1
0.0001	1.0
0.001	10.0
0.01	100.0
0.1	1,000.0
1.0	10,000.0
10.0	100,000.0
100.0	1,000,000.0

General Bibliography

The following references may be helpful to those readers who desire to study the subjects in greater depth.

(1) Bowen, H. H. M., "Trace Elements in Biochemistry," New York, Academic Press, 1966.

(2) Johnson, E., "Trace Elements in Human and Animal Nutrition," 2nd Ed., New York, Academic Press, 1962.

(3) Anonymous, "The supply of trace elements for plants and animals and man," 13 different articles in *Landwirtsch. Forsch. Sonder.*, Germany (1962).

(4) Swaine, D. J., "The Trace Content of Fertilizers," *Commonwealth Bur. Soil Sci. Tech. Commun.*, No. 52.

(5) Wallace, T., "Trace Elements in Plant Nutrition," *J. Roy. Soc. of Arts*, **105** (1957).

(6) "Bibliography of the literature on the minor elements," 4th edition, vol. 2, New York, Chilean Nitrate Educational Bureau, 1951.

(7) Schutte, K. H., "The Biology of Trace Elements: Their Role in Nutrition," Lockwood, C., London, 1964.

(8) Hodgson, J. F., "Chemistry of micronutrient elements in soil," *Advan. Agron.*, vol. **15**, 119–159 (1963).

(9) Mitchell, R. L., "Soil aspects of trace element problems in plants and animals," *J. Res. Agr. Soc.*, **124**, 75–86 (1963).

(10) Swaine, D. J., "Trace Element Content of Soils," *Commonwealth Bur. Soil Sci.*, No. 48, England (1955).

Boron

Peterson, J. R., and McGregor, J. M., "Boron fertilization of corn in Minnesota," *Agron. J.*, **58** (2), (1966).

Berger, K. C., "Boron deficiency, a cause of blank stalks and barren corn ears," *Soil Sci. Soc. Am. Proc.*, **21** (6), 629 (1957).

Albert, L. S., and Wilson, C. M., "Effect of boron on elongation of tomato root tips," *Plant Physiol.*, **36** (2), 244–51 (1961).

Berger, E., "Boron deficiency in apple and pear orchard," "The Deciduous Fruit Grower," vol. **14**, part 2, 49–53, U.S.A. (1964).

Iron

Thomas, G. W., and Coleman, N. T., "Fate of exchangeable iron in acid clay systems," *Soil Sci.*, **97**, 229 (1964).

Perur, N. G., "Effect of iron chlorosis on protein fractions of corn leaf tissue," *Plant Physiol.*, **36** (6), 736–39 (1961).

Hill-Cottingham, D. G., and Lloyd-Jones, C. P., "Analysis of iron chelates in plant extracts," *J. Sci. Food Agric.*, **12**, 69–74, England (1961).

Nagarajah, S., and Ulrich, A., "Iron nutrition of the sugar beet plant in relation to growth, mineral balance, and riboflavin formation," *Soil Sci.*, **96**, 102, 399–407 (1966).

Manganese

Johnson, H. E., et al., "Nutritional limitations and measurements of manganese and molybdenum availability to cotton," *Agr. Chem.*, **21**(8), 18–20, 72 (1966).

McGregor, A. J., and Wilson, G. C. S., "Influence of manganese on the development of potato scab," *Plant Soil*, **XXV**, 3–16 (1966).

Bertrand, D., and Bougault, E., "Choux, manganese et molybdene," *C. R. Acad. Agric. T.*, **50**, (5), 449–451, France (1964).

Page, E. R., "Studies in soil and plant manganese," *Plant Soil*, **XVII** (I), 99–108 (1964).

Zinc

Brown, A. L., et al., "Residual effect of zinc applied to soils," *Soil Sci. Soc. Am. Proc.*, **28**(2) 236–238 (1964).

Hoffman, M., and Samish, R. M., "Control of zinc deficiency in apple," *Israel Agr. Res.*, **16**, 105–114 (1966).

Wallihan, E. F., "Zinc deficiency in the avocado," *Calif. Agr.*, **12**(6), 4–5 (1958).

Hiatt, A. J., and Massey, H. F., "Zinc levels in relation to zinc content and growth of corn," *Agron. J.*, **50** (I), 22 (1958).

Molybdenum

Gupta, U. C., and Mackay, D. C., "Procedure for the determination of exchangeable copper and molybdenum in podzol soils," *Soil Sci.*, **101**, 93–97 (1966).

Kallinis, T. L., "Molybdenum deficiency symptoms in cotton," *Soil Sci. Soc. Am. Proc.*, **31**(4), 507–509 (1967).

Hoover, W. L., and Duren, S. C., "Determination of molybdenum in fertilizers by atomic absorption spectrophotometric method," *J. A.O.A.C.*, **50**(6), 1269–1273 (1967).

Haley, L. E., and Melsted, S.W., "Preliminary studies on molybdenum in Illinois soils," *Soil Sci. Soc. Am. Proc.*, **21**(3), 317 (1957).

Copper

Spencer, W. F., "Effect of copper on yield and uptake of phosphorus and iron by citrus seedlings grown at various phosphorus levels," *Soil Sci.*, **102**, 296–299 (1966).

Larsen, S., "The sorption, desorption and translocation of copper by plants," Agrochimica, X, 190–196 (1966).

Coombs, A. V., "The role of copper sulphate in rice growing," *World Crops*, **16**, 83–84 (1964).

Mitchell, R. L., et al., "Soil copper status and plant uptake," Plant Analysis and Fertilizer Problems, I R.H.O., 249, Paris (1957).

Cobalt

Rana, S. K., "Cobalt status in Quebec soils," *Can. J. Soil Sci.*, **47**(2), 83–88 (1967).

Kubata, J., "Cobalt content of New England soils in relation to cobalt levels in forages for ruminant animals," *Soil Sci. Soc. Am. Proc.*, **28**(2), 246–251 (1964).

Kabata, A., and Beeson, K. C., "Cobalt uptake by plants from cobalt impregnated soil minerals," *Soil Sci. Soc. Am. Proc.*, **25** (2), 125–128 (1961).

Selenium

Lane, J. C., "The determination of selenium in soil and biological materials," *Irish Agr. Res. J.*, **5**, 177–183 (1966).

Davies, E. B., "Uptake of native and applied selenium by pasture species," *New Zealand Agric. Res.*, **9**, 641–652 (1966).

Aluminum

Jones, L. H., "Aluminum uptake and toxicity in plants," *Plant Soil*, **XIII** (4), 297–310 (1961).

INDEX

Adams, S. N., 185
Agric. Res. Council (London), 210
Alabama Agr. Expt Sta., 120
Alfalfa, 118, 226
Alga, (blue green), 181
Alkali disease, 186
Allan, J. E., 236
Allcroft, R., 150
Allison, R. V., 152
Aluminum, 165, 221–222
Ammonium molybdate, 147
Anabaena cylindrica, 181
Anderson, A. J., 134, 139
Anderson, M. S., 196
Andrews, E. D., 203
Anemia, 26, 167
Anions, 15
Aristotle, 3, 20
Arnon, D. I., 8, 12, 30, 134, 219
Ascorbic acid, 141, 152
Askew, H. O., 200
Aspergillus niger, 139, 152
Astragulus bisulcatus, 189
Atomic absorption, 233–236
Atoms, 10
Atriplex vesicaria, 181
Australia, 21, 24, 134, 148, 157, 167, 199, 200, 218
Auxin, 126

Bacon, Francis, 3, 5
Banana, 110
Barclay, C., 218
Bear, F. E., 41, 189, 214, 217
Beath, O. A., 189
Beeson, K. C., 64, 194, 200, 203, 210
Bell, M. C., 169
Berger, K., 48, 119, 139, 143, 158, 175, 176

Bertrand, G., 159
Berzelius, J. J., 5
Blind staggers, 186
Blue chaff disease, 142
Bobko, E., 34
Bokde, S., 115
Bordeaux mixture, 153, 169
Boron, 81–107
    chemistry of, 83
        borax Tronabor, 89
            solubility in water, 84
        content of, in specific plants, 93
        deficiency of, in soils and crops, 86–94
            symptoms of, 87
        occurrence in nature, 83
        use in fertilizers, 88–94
        use on crops, official recommenda-
            tions, 94–99, 103, 106
            alfalfa, 104
            cotton, 99, 103
            orchards, 104, 106
            peanuts
                Fla., 96
                Ga., 94
                Okla., 95
                Southeast, 95, 96
            Soybeans
                Arkansas, 97
Bortels, H., 134, 160, 219
Borys, M. W., 149
Bradford, G. R., 219
Brendenburg, E., 152
Broccoli, 135
Bromine, 207
Brown, A. L., 43, 119, 121, 122
Brownell, P. F., 20
Broyer, T. C., 20, 172
Brun, T. S., 155
Bukovac, M. J., 43
Burgess, W. D., Jr. 237

Calderbank, G., 210
California, 195, 218, 219
Camp, A. F., 107
Carbon dioxide, 6
Carbonate zinc, 117
Carlton, A. B., 20
Cations, 15
Cauliflower, 134, 141
Caventon, J. B., 16
Chalcopyrite, 153
Chalcosite, 153
Chandler, W. H., 107
Chapman, H. L., 169
Chelates, 50–53
Chemical fertilizers, 31
Chenery, E. M., 222
Chernozem soil, 25
Childers, N. F., 149
Chilean nitrate, 207, 210
Chlorine 20, 172–179
    chemistry of, 173–174
    in animal nutrition, 177–179
    in plants, 174–177
    sources of, 174
Chlorophyll, 16, 152
Chlorosis, 42
Chromium, 21
Citrus leaf spot, 147
Clemson University, 234
Climax Molybdenum, 146
Coastal Bermuda, 123
Coastal Plain soils, 156
Cobalamine, 201
Cobalt, 22, 197–206
    as pasture fertilizer, 204
    chemistry of, 199
    corrective methods, 203
    deficiency disease of, 199–200
    drenching with, 206
    function ot, 201
    in plants, 202–203
    in salt licks, 206
    in soils, 202
    occurrence, 197–198
    with vitamin B_{12}, 200–201
Cobalt glance, 199
Cobalt speiss, 199
Cobaltous chloride, 199, 204

Cobaltous sulfate, 199, 204
Colorado soil test, 234
Colorado State University, 199
Columella, 5
Cook, F. C., 88
Copper, 151–171
    as fungicide, 169
    chemistry of, 152–154
    comparative toxicity of, 169–170
    content in wheat, 163
    copper compounds, 153–154
    copper sulfate pentahydrate, 153–154
    deficiency symptoms of, 161–163
    fertilization, 162
    functions of, 167
    general bibliography, 170–171
    in animals, 167
    in enzymes, 159–160
    in green plants, 160–161
    in other diseases, 168–169
    in soils, 154–158
    interrelationships, other trace ele-
        ments, 165–167
    tract, 168
Corn, composition of, 27
Covellite, 153
"Crinkle leaf," 74
Cunningham, H. Gordon, 229

David, D. J., 234
Davy, H., 5
Dawson, C. R., 159
De Groot, T., 178
Delphinidine diglycoside, 222
Denmark, 157
de Ong, E. R., 169
Depletion factors, 35
Diagnosis, visual, 34
Dick, A. T., 150
Dietz, C., 209
Dijkshoorn, W., 174
Dithizone-extractable Zn, 122
Dixon, J. K., 200
Doney, R. C., 43
Drake, C., 190
Drenching, 206
Duncan, D. B., 235
Dundas, J., 185

du Toit, J. L., 125

Easton, F. M., 172
Eddings, J. L., 43
Electrons, 12
Elements, chemical, 11, 31
Ellis, B. A., 121
Emulsions, 13, 14
Enzootic inarasmus, 200, 218
Enzymes, 23
  ascorbic acid oxidase, 160
  carbonic anhydrase, 111
  laccase, 160
  nitrate reduction, 140, 141, 149
  tyrosinase, 159
Essentiality, 12
Evans, H. E., 124
Exanthema, 163
Exudative diathesis, 188

"Factor 3", 188
Fearon, W. R., 77, 221
Ferguson, J. K., 181
Ferromolybdate, 136
Ferrous sulfate solution, 50
Filmer, J. F., 200
Finland, 118
Flax, 160
Fleming, G. A., 174, 179
Florida peats, 152, 155
Florida soils, 227
  citrus growers, 146
Fluorides, 25
Fluorine, 213–216
  in animals, 215
  in plants, 214–215
  sources: cryolite, fluorapatite, fluorspar, micas, topaz, tourmaline, 213–214
Fluorosis, 215
Folz, C. M., 187, 194
France, 207
Freney, J. R., 148
Fresno County, 50
Fricke, E. F., 142
Frits, 118, 229, 234
Fudge, J. F., 210
Fuller, G., 195

Galileo, 3
Garber, R. J., 189
Gatehouse, B. M., 236
Genetics, crop yields, 29
Geneva (N.Y.) Expt. Sta., 185
Georgia Agr. Expt. Sta., 147, 148
Gerloff, G. C., 183
Gilbert, F. A., 156, 158, 161, 207
Gladstone, J. S., 124
Glass, B., 150
Gnadinger, C. B., 194
Godden, W., 211
Goiter, 208, 209, 211
Grant, A. B., 190, 196
Grassland herbage, 123
Gray speck disease, 165
Great Britain, 202
Green, H. H., 161

Haemoglobin, 167
Halogens, 207
Hammer, H. E., 234
Hardison, W. A., 203
Harmer, P. M., 156
Harmsen, H. E., 183
Hartley, W. J., 190, 196
Hawaiian Islands, 146
Helmont, J. B. von, 5
Henkins, C. H., 183
Herbage plants, 32
Hewitt, E. J., 141
Hill, R., 152
Hoagland, D. R., 8, 19, 107, 125
Hodgson, J. F., 123
Hogan, A. G., 184
Holland, 151
Holmes, R. S., 155
Holmes, W. J., 183
Hopkirk, C. S. M., 212
Horecker, B. L., 222
Hudig, J., 151
Hunger Signs, 37
"Husbandry", by Columella, 5
Hybrids, corn, 29
Hydrangea, 222
Hydrofluoric acid, 215
Hydrogen selenide, 188

Ingenhousz, Jan., 4
Iodine, 207–212
    goiter, 211–212
    in animal nutrition, 210–211
    occurrence, 207
        in plants, 209–210
        in soils, 208–209
Iodine Educational Bu., 208
Ions, 13, 15
Iron, 39–57
    chemical relationships, 40
    corrective measures, 49–50
        in animal diet, 55
        in piglet anemia, 55
    deficiency of, 45
        diagnosis, 48
        visual symptoms of, 46
    in leaves, 43, 44
    in soils, 40–41
    in various plant products, 45
    minerals: biotite, carbonate, chlorite,
        hematite, hornblende, magne-
        tite, pyrite, siderite, 39
Iron dextran, 56
Iron porphyrin, 152

Jacob, A., 86
Javallier, M., 107
Johnson, C. M., 20
Jones, E. W., 142
Jones, J., 122

Kelp, 207
Kemp, A., 175
Knop, W., 8
Krantz, B. A., 50, 121, 122

Lakanen, Esto, 118
Lanning, F. C., 220
Lavoisier, Antoine, L., 4
Leeper, G. W., 61
Leitch, I., 209
Lemon leaves, 139
Leonard, C. D., 52
Lettuce, 140
Lewis, G., 150
Liebig, J. von, 28
Limestone, 25, 118

Lindsey, W. L., 120
Linnaeite, 199
Lipman, C. B., 20, 107, 160, 221
Lithium, 218, 219
"Little Leaf", 125, 126
Loneragan, J. F., 124
Lucas, R. E., 161

McCallan, S. E. A., 169
McCollum, E. V., 222
McConnell, P., 41
McElroy, W. D., 21, 150
McHargue, J. S., 20, 88
McKinney, G., 20, 160
McLean, J. W., 188
McMurtrey, J. E. Jr., 114, 156, 202, 221
Madison, T. C., 186
Magistad, O. C., 222
Malachite, 153
Mallette, M. F., 159, 178, 184
Manganese, 58–81
    catalyst, 58
    chemical relationships, 59
        amount in soil, 61
        as manganic ion, 59
        as manganous ion, 59
        methods to determine amount, 60
    deficiency of
        in plants, 68–72
            in alfalfa, 72
            in beans, 74
            in carrot and lettuce, 73
            in cauliflower, 73
            in potato, 73
            in soybeans, 72
            in tobacco, 74
        in small grains, 74
        in soils, 65
            corrective measures, 67
            symptoms of, 66
    in animal metabolism, 74, 75, 77
    in fertilizer, 79
    reactions of
        in plants, 64
        in soils, 63
    removal of, in crops, 65
    toxic dose of, 77
    toxicity symptoms, 74

Manure salts, 36
Market, size of, 231
Martin, P. E., 121, 122
Mayer, J. R., 16
Maynard, L. A., 221
Meristem cells, 15
Meyer, C., 151
Michigan, 185
Michigan Agr. Expt. Sta., 121
Micronutrients, 23
Milk, 167
Millikan, C. R., 140
Mitchell, R. L., 26, 111, 124, 155, 202, 205, 217
Moghe, V. B., 110
Molecules, 12
Molybdenite, 136
Molybdenum, 133–150
   chemical relationships of, 136
   deficiency of, 141–143
     correction of, 146–147
     symptoms of, 141–143
   experiments involving, 147–149
   general references on, 149, 150
   in grassland herbage, 144–146
   in plants, 138, 139
   in soils, 136, 138
   primary minerals of, 136
   toxicity of, to animals, 143–144
Morton Mains disease, 199, 218
Mortredt, J. J., 120
Moschler, W. W., 118
Mulder, E. G., 159
Mulder, G. J., 5

Natal (S. Africa) Cane Belt, 125
Nelson, E. M., 195
Nemec, A., 220
Neutrons, 12
New Zealand, 134, 135, 147, 189, 190, 192, 195, 199, 204, 208, 218
Nickel, 22, 217, 218
Nierman, J. L., 184
Nikolic, S., 25
Ninety Mile Plain, 157
Norris, L. C., 77
North Dakota, 114
Nutrition, mineral, of plants, 26

NuZn, 123

Oats, 142
Oetting, W., 85
Ohio Agr. Res. Center, 122
Onions, 163
Oostendorp, D., 183
Oota, M., 220
Orange leaves, 142
Oregon, 192
Organic fungicides, 170
Orr, J. B., 175, 209
Oxygen, 6
Ozanne, P. G., 173

Page, N. R., 234, 237
Palissy, B., 5
Pang, T. S., 114
Parakeratosis, 127
Patterson, R. M., 121
Peaches, 116, 117, 126
Peat soils, 24, 155, 163, 164
Pecan rosette, 107
Pecans, 126
Pelletier, J., 116
Pennsylvania State College, 29
Perkin Elmer "Model 303", 233, 235
Perosis, 77
Peterson, N. K., 33
Phosphate, 24
Photosynthesis, 4, 6, 16
pH, soils, 30, 158
Pining disease, 200
Piper, C. S., 135
Podzolic soil, 25
Polstorff, L., 7
Polyflavenoid, 229
Potassium, 182
Powellite, 136
Powers, W. L., 114
Pratt, P. F., 48
Price, N. O., 203
Priestley, J., 4, 6
Primary producers, 230
Protons, 12
Purdue University, 56
Purvis, E. R., 33

Radicals, 13
Rahman, H., 174
Raleigh, G. J., 220
Rayplex Zn, 121
Reclamation disease, 24, 151, 161
Reith, J. W. S., 124, 184
Rhizobium, 201
Rickets, 56
Roach, W. A., 218
Robinson, W. O., 114, 137, 202, 221
Roots, 16
Rosenfeld, I., 189
Rosetting, 125, 126
Rothamsted Expt. Sta., 19, 185
Rukuhia Soil Res. Sta., 189
Russell, F. C., 61
Russell, E. J., 185
Rutgers University, 64, 149, 217

Sachs, J., 8
Salt Lick, 206
"Salt sick", 168
Salter, R. M., 61
San Diego County, 50
Saussure, Theo. de, 4, 7, 16
Scarseth, G. D., 61
Scenedesmus obliguus, 219
Scharrer, K., 85
Scheele, Carl, 6
Schwarz, K., 187, 188, 194
Scotland, 202, 217
Scott, M. L., 188, 194
Selenic acid, 188
Selenious acid, 188
Selenium, 186–196
    as insecticide, 194–195
    chemical analysis of, 195, 196
    chemistry of, 188
    health hazards, 195
    preventive beneficial treatment, 190–
        194
    research on, 188, 189
    supplementation of, 189–190
Selocide, 194
Sequestrene Zn, 123
Serviss, G. H., 61, 232
Shot gun mixture, 227
Silicon; silica, 219, 220, 221

Silicon fluoride, 215
Smith, W. S., 152
Sodium, 180–185
    benefits from use, 181, 182
    effect on soil structure, 182, 183
    in animals, 184
    in plants, 183, 184
    occurrence of, 180, 181
    use as fertilizer, 184, 185
Sodium chloride, 173, 178
Sodium iodate, 208
Sodium nitrate, 183, 207
Sodium selenate, 192, 195
Sodium tetraborate, 89
Soil Science Society, 143
Soil test, Colorado, 234
Solubor, 105
Solutions, definition of, 13
Sommer, A. L., 151, 160
Soybeans, 126
Spectrophotometer, 233
Sphalerite, 108
Sprengel, Carl, 28
Steinbeck, O., 85
Steinberg, R. A., 139, 160
Stewart, I., 52
Stiles, W., 150
Stotz, E. H., 160
Stout, P. R., 19, 20, 134, 176
Strontium, 25
Sugar beet, 177, 185
Sugar cane, 124, 164
Sulfur, 36
Sullivan, J. T., 189
Sunflower, 160
Superphosphate, 36, 79, 147
Swaine, D. J., 171
"Swingback", 168

Tasmania, 142
Taylor, A. W., 220
Taylor, G. G., 195
Teakle, L. J. H., 157
Teart, 143
Tenorite, 153
Terman, G. L., 120
Thacker, E. J., 200
Thaer, A. D., 5

Thatcher, R. W., 111
Thyroids, 212
Thyroxine, 211
Tisdale, W. B., 171
Titanium, 22
Tobacco, 125, 156
Todd, W. R., 194
Tomato, 135, 141, 160, 163, 177
Toui Cheng, 125
Tourmaline, 87
Tulare County, 50
Tung, 107, 108, 127
TVA, 93, 120

Vacuolar sap, 17
Vanadium, 21, 22, 219
Vinquish, 200
Virginia Agr. Expt. Sta., 119
Visual diagnosis, 34
Vitamin B$_{12}$, 22, 23, 198, 201
Vitamin C, 160
Vitamin D, 26
Vitamin E, 188, 193

Wallace, T., 61, 156
Walsh, A., 236
Wanaka, (N. J.), 212
Ward, G. M., 125
Warington, K., 19, 20
Washington Agr. Expt. Sta., 122, 148, 149
Water Culture, 8
Watkinson, J. H., 196
Wear, J. I., 119, 121
Wehrman, J., 202
Weichelsbaum, T. E., 194
Wessel, G., 219
Wheat, 161
Whiptail, 134, 135
Whitehead, D. C., 146
White bud, 115

White muscular disease, 187, 190
Whitney, R. S., 118
Wiegmann, F. A., 7
Wilcoxon, F., 169
Wisconsin, 158, 226
Wittwer, S. H., 43
Wolf, E., 192
Wood, J. G., 20
Woodward, J., 6
Wulfenite, 136
Wyoming, 189

Yeast (brewer's), 188
Yellow spot, 143

Zinc, 20, 107–132, 229
    chemical and physical characteristics, 20, 108, 109
    conversion chart, 132
    deficiency, 119–127
        in sugar cane, 124
        in tobacco, 125
        major symptoms
            in animals, 127
            in plants, 126
    in grassland herbage, 123
    in plants, 110, 112
        function of, 111
    in soil, 114
        reactions in, 118
    recommendations by states, 128–131
    solubility, 109, 116
    sources for fertilizer uses, 109
Zinc ammonium phosphate, 118
Zinc carbonate, 123
Zinc chelate (Zn EDTA), 118, 121, HEEDTA, 121
Zinc NTA, 121
Zinc oxide, 109
Zinc sultate, 118
Zircon, 25